Peptides and Proteins

Cambridge Chemistry Texts

Peptides and Proteins

D. T. ELMORE

Department of Biochemistry,
The Queen's University of Belfast

CAMBRIDGE
at the University Press, 1968

Published by the Syndics of the Cambridge University Press
Bentley House, 200 Euston Road, London, N.W.1
American Branch: 32 East 57th Street, New York, N.Y.10022

Library of Congress Catalogue Card Number: 68-21392

Standard Book Number: 521 07107 0

Printed in Great Britain
at the University Printing House, Cambridge
(Brooke Crutchley, University Printer)

Contents

Preface

The present volume is designed to give the advanced undergraduate student in chemistry an insight into the field of proteins. Most of the book is devoted to the structural and synthetic aspects of protein chemistry, but this would hardly give a contemporary and balanced picture if it were presented in a biological vacuum. Consequently, such topics as protein biosynthesis and the relation between chemical structure and biological activity have been included. The artificial line of demarcation between chemistry and biochemistry has become more nebulous in the protein field in recent years and it is hoped that this book will in some small way help to erase it entirely. If this is to come to pass, not only will students of chemistry be expected to read some biochemistry, but biochemists will have to become better chemists. It is hoped, therefore, that students of biochemistry who use the book will overlook the inevitable condensation of the more biological aspects of proteins and will be encouraged to think of enzymes and other proteins as molecules whose biological activity is defined by their chemical structure and molecular shape.

Selectivity has had to be exercised in choosing examples of blocking groups for peptide synthesis, methods of forming the peptide bond and methods of degrading peptides and proteins. Those synthetic and degradative methods described are widely used, appear to be promising, or are interesting from a mechanistic point of view. Further, no attempt has been made to derive mathematical equations which are required for the study of macromolecular properties and structure of proteins. Physical chemists will no doubt protest at this emasculated approach to the subject, but two considerations were of overriding importance. First, physical techniques provide valuable information about the structure and properties of proteins; it is the information, its interpretation and relationship to knowledge from other sources, rather than the techniques themselves, which concern us here. Secondly, it was essential to keep the book down to a size and price which might attract the average undergraduate.

Many other books on proteins have been written, some extensive and some of more modest dimensions. Looking at these, one wonders if some other technique might not have been squeezed in or if some material included in the present book might not have been condensed or omitted. Better authors than the present must have wrestled with this problem and one's doubts are to some extent allayed by Samuel Butler's remark: 'When a man is in doubt about this or that in his writing, it will often guide him if he asks himself how it will tell a hundred years hence.' With a subject which is developing as fast as the field of peptides and proteins, the answer is painfully obvious.

Belfast
August 1967

D.T.E.

Nomenclature

A peptide is a compound in which amino acids are covalently linked together through secondary amide or peptide bonds (—CO.NH—). The term protein, derived from the Greek word *proteios* ('primary'), was coined before anything was known about the chemical structures of these natural macromolecules. It is conventional to retain this term for large molecules and to use the term peptide for molecules which are built up from only a small number of amino acid residues. Nature, however, is not influenced by terminological conventions and the molecular sizes of natural peptides and proteins cover a fairly continuous spectrum. It is debatable, therefore, whether adrenocorticotropic hormone, which is built up from thirty-nine amino acid residues, is a large peptide or a small protein. Use of the terms oligopeptide and polypeptide are similarly vague; in addition, the latter term has been reserved by some authors for artificial macromolecules built up from only one type of amino acid residue. It is probably too late to hope for universal acceptance of a more systematic nomenclature. Fortunately, the meanings of the various terms are usually quite clear from the context in which they occur. In particular cases, the size of a molecule may be specified by a prefix, e.g. undecapeptide. In sport, an analogous situation exists with the word team. This term applies to any group who play together against another team or teams and might range from two (e.g. tennis) to a very large number (e.g. a national team in an athletics or swimming fixture). In particular cases, a prefix in the form of an adjectival noun (e.g. cricket) defines the size of the team. In the protein field, other less precise terms such as peptone, primary proteose and secondary proteose, which have been used as a vague indication of molecular size, seem to have dropped out of use almost completely. They will not be resurrected here.

It is conventional to write the structure of a peptide with a free amino acid group at the left-hand side and a free carboxyl group on the right-hand side. Peptides are named by regarding them as acylated derivatives of the *last* amino acid residue. The names of the acylating residues are then derived frcm the trivial name of the amino acid by replacing the ending, usually -ine, by -yl. For example, $NH_2.CH_2.CO.NH.CH(CH_3).CO.OH$ is named glycylalanine because it can be regarded as an *N*-acylated derivative of alanine.

Throughout this book, the abbreviations recommended by the International Union of Pure and Applied Chemistry [*IUPAC* Information Bulletin, no. 20 (1963)] are used (table 2.1). The configuration is assumed to be L-unless otherwise stated. In sequences, a hyphen between two abbreviations indicates that the amino acids are joined through a peptide linkage, e.g. Gly-Ala represents $NH_2.CH_2.CO.NH.CH(CH_3)CO.OH$. Where abbreviations are separated by commas and enclosed in parenthesis, the sequence is unknown.

The generally accepted method of classification of proteins, except those which function as enzymes, dates back to 1907 and is most unsatisfactory. There are two broad classes: unconjugated proteins, which contain only amino acids, and conjugated proteins, which contain additional types of groups. Unconjugated proteins are then classified in the main according to their solubility properties. *Albumins* are soluble in water and dilute solutions of salts; *globulins* are sparingly soluble in water but are soluble in dilute solutions of salts; *prolamines* are insoluble in water but are soluble in 50–90 per cent aqueous ethanol; *glutelins* are insoluble in all the foregoing solvents but are soluble in dilute solutions of acids or bases; *scleroproteins* are insoluble in most solvents; *protamines* contain a high proportion of arginine and are strongly basic; *histones* are also very basic but they contain a greater variety of amino acids than protamines. Conjugated proteins are classified according to the kind of group they contain: *nucleoproteins* are complexes of nucleic acids and basic proteins such as protamines and histones; *lipoproteins* and *glycoproteins* contain respectively lipids and carbohydrates linked to the polypeptide; *chromoproteins* contain a ligand which absorbs light in the visible part of the spectrum.

A modern classification of proteins might be based on structure (e.g. molecular size, amino acid composition, helical content) and function (e.g. skeletal proteins, enzymes, hormones, antibodies). The only step so far taken is the recent classification of enzymes by the International Union of Biochemistry. Six main classes are recognised depending on the type of reaction which is catalysed. *Oxidoreductases* catalyse redox reactions; *transferases* catalyse reactions such as transamination and transphosphorylations; *hydrolases* include enzymes which hydrolyse esters, glycosides and peptide and other amide bonds; *lyases* either remove groups from their substrates leaving double bonds (e.g. decarboxylation, β-elimination) or conversely add groups to double bonds; *isomerases* catalyse racemization, epimerization, cis-trans isomerizations and the interconversion of aldoses and ketoses; *ligases* or *synthetases* catalyse the formation of C—O, C—N or C—C bonds with the concomitant breakdown of a nucleoside-5′ triphosphate to a mono- or diphosphate. The systematic name of an enzyme consists of two parts. The first consists of the substrate or substrates; the second part, ending in '-ase' indicates the nature of the reaction catalysed. For example, *N*-acetylmuramide glycanohydrolase is the systematic name for lysozyme (see p. 69).

1. Separation and isolation of peptides and proteins

1.1. Extraction of proteins from cells. Although proteins consist of amino acids which are linked together by covalent bonds, their macromolecular structure depends on the presence of additional weak linkages such as hydrogen bonds, hydrophobic bonds and electrovalent bonds (see chapter 3). Since the biological activity of a protein depends on the integrity of its molecular conformation, methods of isolation must be sufficiently mild not to disrupt these weak bonds and cause denaturation (see §**3.10**). On the other hand, proteins frequently occur in the form of complexes with nucleic acids, lipids or polysaccharides and these may form insoluble particles. Determination of the optimum conditions for extraction of proteins is semi-empirical and may be extremely laborious. In general, proteins are sensitive to heat, extremes of pH and the presence of high concentrations of organic solvents or detergents. In addition, it is often found that proteins are more stable in concentrated than in dilute solution, possibly because the molecules tend to associate into dimers or larger units.

Before extracting proteins from tissues, cellular structure must be destroyed and various techniques are available for this purpose. The tissue may be macerated, usually in the presence of an aqueous diluent, in a blendor fitted with high-speed rotating knives. For small quantities the Potter–Elvehjem homogenizer, consisting of a rotating plastic pestle in a glass tube which can be raised and lowered, has the advantage that subcellular structures survive disintegration. If subcellular structures such as mitochondria are required intact, homogenization is often performed in 0·25 M-sucrose solution and the particular fractions are separated by differential centrifugation. Other procedures for disrupting the cellular structure include alternate freezing and thawing, exposure to ultrasonic vibrations or careful drying to a powder by treatment with acetone at a low temperature.

Extraction of proteins from disintegrated cells or isolated particulate fractions is usually achieved with water or dilute solutions of salts such as sodium chloride. The extraction is frequently carried out at a controlled pH by using buffered solutions. Some proteins (globulins) are insoluble in water but soluble in dilute salt solutions. Acetone-dried powders may be extracted in the same manner as is used for moist, disintegrated tissue.

1.2. Precipitation of proteins. Having crudely extracted a protein from tissue material, it is common practice to attempt to precipitate it preferentially from solution. Factors which influence the solubility of proteins include temperature, pH, ionic strength and the presence of organic solvents. The effect of temperature depends on whether the heat of solution is positive or negative. In general, however, control of protein solubility by adjustment of temperature is not widely used because proteins are usually denatured rapidly at elevated temperatures. Denatured proteins are generally much less soluble than native proteins and occasionally unwanted protein can be precipitated by selective thermal denaturation leaving the desired protein in solution.

Proteins have a minimum solubility near their isoelectric point. The isoelectric point may be defined, for present purposes, as the pH at which the numbers of positive and negative charges on the molecule are equal. The use of this phenomenon for the separation of proteins is illustrated in fig. 1.1. Proteins A and B have the same isoelectric point, but only the latter is likely to precipitate at this pH. Protein C, like protein B, has a very low solubility at its isoelectric point. Since protein C is quite soluble at the isoelectric point of protein B and *vice versa*, the two proteins could be precipitated separately by adjustment of the pH of the solution first to the isoelectric point of one of the components and, after centrifugation, to the isoelectric point of the other protein.

As mentioned above, globulins are insoluble in water, but are 'salted-in' by low concentrations of salts. At high salt concentrations, however, proteins are generally precipitated, a phenomenon known as 'salting-out'. Salts containing multivalent ions, such as sulphate or phosphate, are more effective than salts containing monovalent ions (fig. 1.2). For the 'salting-out' process in concentrated electrolytes, the solubility, S, and ionic strength, I, are related by the equation:

$$\log S = \beta - KI,$$

where K is the 'salting-out' constant. The value of K varies with the nature of the protein and salt but is independent of tempera-

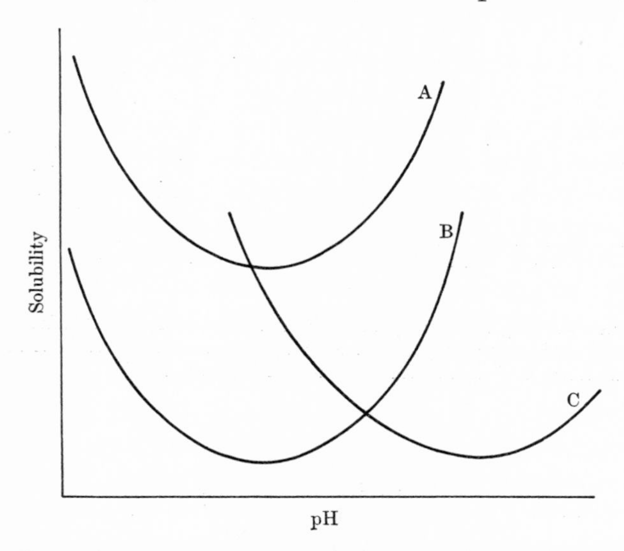

Fig. 1.1. The variation of solubility of a protein with pH. Proteins A and B have the same isoelectric point, but only the latter precipitates there. Protein C, at its own more alkaline isoelectric point, has a low solubility, but is quite soluble at the isoelectric point of proteins A and B.

ture and pH. On the other hand, the value of β is markedly dependent on temperature and pH. Typical plots of $\log S$ versus I are shown in fig. 1.3. 'Salting-out' is widely used as a means of separating proteins and is clearly most selective when K is high. Ammonium and sodium sulphates are commonly used. The method is particularly useful in the early stages for isolating a protein, since it is applicable to large quantities.

Water-miscible organic solvents, by lowering the dielectric constant, also precipitate proteins. The danger of denaturation limits the range of useful solvents and the temperature must usually be kept below 0 °C. The Cohn method for the fractionation of plasma

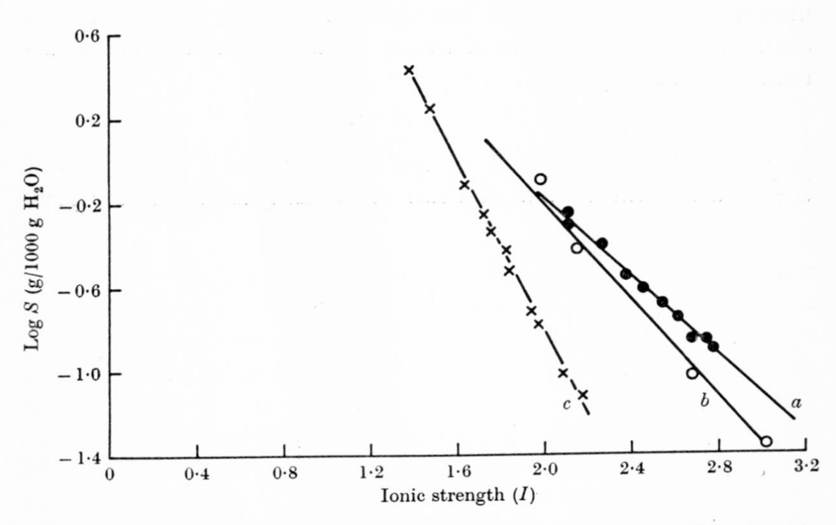

Fig. 1.2. The solubility of fibrinogen in different salt solutions: (*a*) sodium chloride at pH 5·8; (*b*) ammonium sulphate at pH 6·0; (*c*) potassium phosphate at pH 6·6. [Data taken from M. Florkin (1930). *J. biol. Chem.* **87,** 629.]

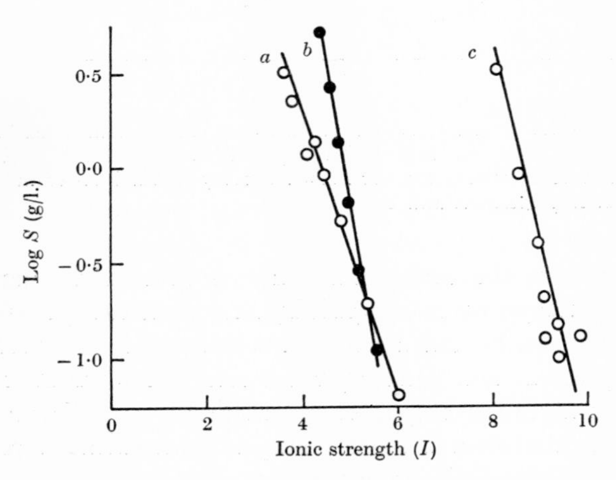

Fig. 1.3. The solubility of proteins in ammonium sulphate solutions: (*a*) haemoglobin; (*b*) pseudoglobulin; (*c*) myoglobin. [E. J. Cohn & J. T. Edsall (1943). *Proteins, Amino Acids and Peptides,* p. 602. New York: Reinhold Publishing Corporation.]

proteins by precipitation with aqueous ethanol is a good example of the application of this technique (Cohn, Strong, Hughes, Mulford, Ashworth, Melin & Taylor, 1946).

1.3. Dialysis. When two solutions of different concentration are separated by a permeable membrane such as cellophane, inorganic ions and small molecules, but not macromolecules, can diffuse through the membrane. This process, known as dialysis, has long been used for the removal of salts after a 'salting-out' step in the isolation of a protein. The protein solution is placed in a length of cellophane tubing, knotted at one end, and suspended in a stream of running water. Salts and other small solutes diffuse through the membrane and the nondialysable protein can be recovered from the solution left in the dialysis sac. Dialysis membranes probably function as molecular sieves, so that the rate of diffusion through the membrane will depend on pore size and the molecular dimensions of the solute. Dialysis has been used with some success for the fractionation of small polypeptides such as the peptide antibiotics. In particular, the counter-current principle (see p. 10) has been applied (Craig, 1960*a*). The process of dialysis can be accelerated by using a membrane area which is large relative to the volume of solution to be dialysed.

1.4. Freeze-drying. Proteins cannot be recovered in the native state from aqueous solution by evaporation of liquid even when reduced pressures are used to lower the boiling-point. Instead, the aqueous solution is frozen and the ice is sublimed at very low pressures using receivers cooled in acetone–solid carbon dioxide mixture or liquid nitrogen to trap the ice and keep its vapour pressure low. The solution must be kept frozen throughout the process and the protein is usually recovered as a fluffy, hygroscopic powder.

1.5. Column chromatography. It is not possible to deal in detail here with the general principles and practice of chromatography, and a number of textbooks are recommended for consultation (Lederer & Lederer, 1957; Morris & Morris, 1963; James & Morris, 1964). Only those techniques which have found major application for the separation of proteins will be discussed here.

Separation of mixtures by column chromatography is achieved by applying a solution of the mixture as a narrow band to a column of a solid stationary phase followed by elution with solvents, whose composition is altered either continuously (gradient elution) or discontinuously (stepwise elution). The rate of elution should be slow enough for equilibrium to be established at each point. The physical factors which determine the rate at which a solute travels down the column include adsorption due to hydrogen-bonding or hydrophobic bonding, ion-exchange, molecular sieving, and partition between partially miscible phases. It is probable that more than one factor is operative in most cases. Theoretically, the concentration of solute throughout a band should follow a Gaussian distribution when plotted against the volume of eluting solvent. For a variety of reasons, this behaviour is frequently not realized in practice and bands tend to 'tail' and may not separate completely. Gradient elution tends to overcome 'tailing' and various devices, such as the 'Varigrad' (Petersen & Sober, 1960) have been described for producing linear, concave and convex concentration-gradient profiles. For the chromatography of macromolecules such as proteins, gradient elution offers an additional advantage, since fractions obtained by stepwise elution may contain the same material in different peaks.

Chromatography of proteins on columns of calcium phosphate has been thoroughly investigated by Tiselius. The behaviour of calcium phosphate as an adsorbent for proteins depends critically on its method of preparation, since both physical and chemical changes in the adsorbent readily occur. Tricalcium phosphate, brushite [$CaHPO_4 . 2H_2O$] and hydroxylapatite [$Ca_5(PO_4)_3 . OH$] have all been extensively used. Proteins are bound to calcium phosphate by ionic forces and elution is effected by changes in ionic strength and pH. The affinity of calcium ions for some proteins may be especially important. An example of the use of a hydroxylapatite column for the separation of serum proteins is shown in fig. 1.4.

Molecules may be separated by virtue of their molecular dimensions using the technique of gel-filtration or exclusion chromatography. A chromatographic support is chosen for its ability to behave as a molecular sieve; molecules penetrate the particles of the support to varying extents depending on their molecular

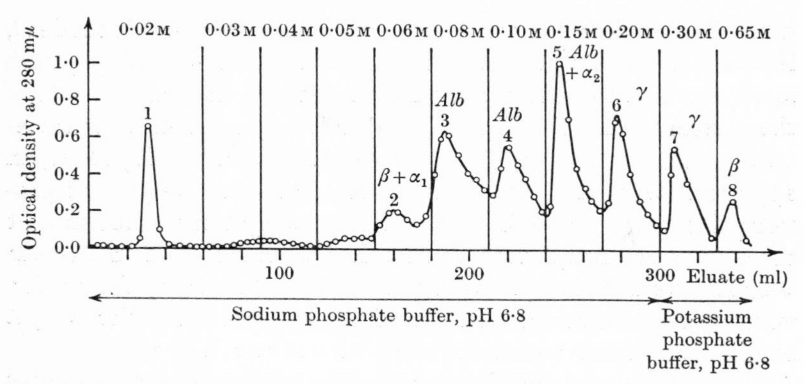

Fig. 1.4. The chromatography of human serum proteins on a hydroxylapatite column. α, β and γ refer to globulin fractions; *Alb* represents albumin. [From S. Hjerten (1959). *Biochim. Biophys. Acta* **31**, 216.]

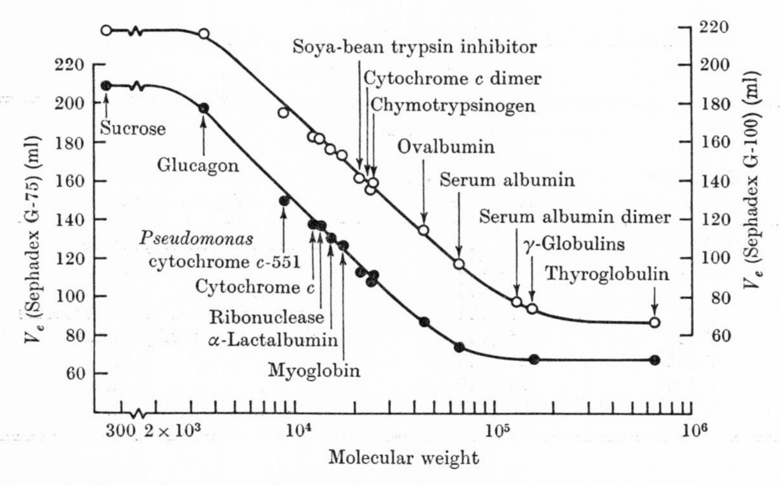

Fig. 1.5. Plots of elution volumes, V_e, against log(molecular weight) for proteins on Sephadex G-75 (●) and G-100 (○) columns [2·4 cm × 50 cm; equilibrated with 0·05 M-tris hydrochloride buffer, pH 7·5, containing KCl (0·1 M)]. [P. Andrews (1964). *Biochem. J.* **91**, 222.]

dimensions. In an extreme case, if a solution containing a mixture of an inorganic salt and a protein is passed down such a column, the ions can penetrate freely into the column while the protein, if the molecules are sufficiently large, will be completely excluded and will travel with the solvent front. Obviously, gel filtration can be used in place of dialysis for removing salts from protein solutions. Cross-linked dextran, commercially available as 'Sephadex', is a suitable support for gel filtration. If a mixture of polypeptides is applied to a column of cross-linked dextran, the components of the mixture will be eluted in the decreasing order of their molecular weights. In fact, gel filtration has been employed as an approximate method for determining molecular weights of proteins. Using proteins of known molecular weight, it was found that the elution volume plotted against the logarithms of the molecular weight gave a smooth curve (fig. 1.5). By determining the elution volume of a protein on a calibrated column, the molecular weight can be determined by interpolation on the standard curve.

The usefulness of gel filtration for the fractionation of mixtures of proteins or peptides has been extended by the availability of chromatographic supports with differing porosities (table 1.1).

TABLE 1.1. *Exclusion limits for gel filtration on 'Sephadex' columns*

'Sephadex' grade	Approximate molecular weight of smallest excluded molecule	Applications
G 10	700	Fractionation of peptides
G 15	1,500	" "
G 25	5,000	" "
G 50	10,000	Fractionation of small proteins
G 75	50,000	Fractionation of proteins
G 100	100,000	" "
G 150	150,000	" "
G 200	200,000	" "

Separation of small peptides by ion-exchange chromatography is similar to that employed for the quantative analysis of amino acids (see p. 23), and the same mechanized procedure can be used. The relationship between the elution volume and the state of charge of the peptide is reasonably well understood. Proteins, on the other hand, behave less predictably on ion-exchange resins. Since protein molecules may contain many charged groups, it is not certain that equilibrium is established at each point on the column. Secondly, there are also steric factors to be considered. Not only may the protein molecules arrive at the surface of the resin with different orientations, but some of the charged groups in the protein molecule may be in the interior of the molecule and be unable to form ionic bonds. Thirdly, the protein molecules may be unable to penetrate the resin particles readily and molecular sieving may occur in addition to ion-exchange. Fourthly, it is

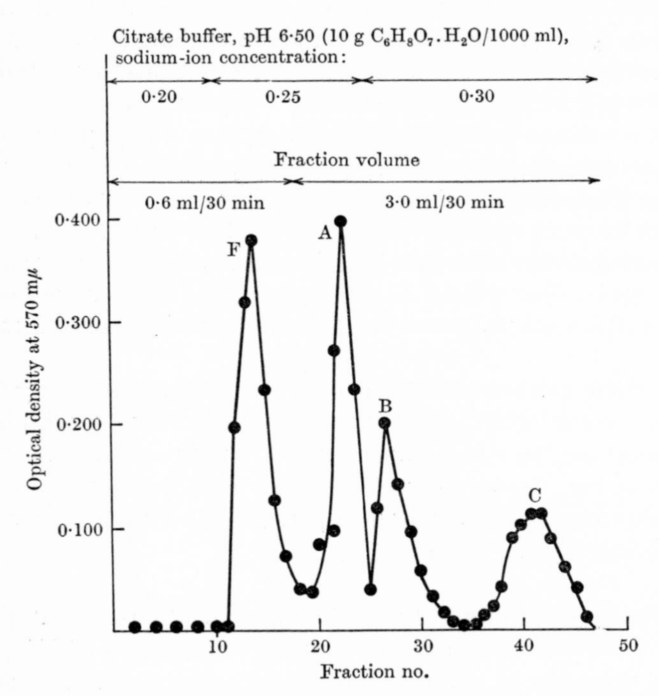

Fig. 1.6. Chromatography of four different kinds of human haemoglobin on Amberlite IRC-50. [H. K. Prins & T. H. J. Hinsman (1955). *Nature*, **175**, 903.]

likely that hydrogen-bonding and hydrophobic bonding play a part in binding proteins to ion-exchange resins. Finally, change of ionic strength and/or pH during elution chromatography may produce subtle changes of molecular conformation thereby altering the number and steric relationship of groups in the protein molecule which are available for binding. In spite of the nebulous interpretation of the behaviour of proteins on ion-exchange resins, this chromatographic technique is extremely useful. There are many examples of successful separations of very closely related proteins (fig. 1.6). Probably, the most widely used synthetic resin ion-exchanger is Amberlite IRC–50, which has carboxyl groups as ionizable functions.

Ion exchangers have also been made by chemically modifying cellulose. Substituents such as carboxymethyl (CM) and *N*-diethyl-aminoethyl (DEAE) groups have been introduced into cellulose and provide respectively, cation- and anion-exchange resins. The open structure of these ion-exchangers allows proteins to penetrate readily. Several other substituted cellulose derivatives have been made and are commercially available. Examples of the application of these ion-exchangers to the fractionation of protein mixtures are legion. For example, human serum has been separated by chromatography on a column of DEAE-cellulose into a large number of fractions some of which could not be resolved by electrophoresis. Again, four closely related trypsin inhibitors have been isolated by chromatography on DEAE-cellulose of extracts of lima beans.

Similar chemical modifications have been applied recently to 'Sephadex' cross-linked dextrans to produce chromatographic adsorbents which have the properties of both ion-exchangers and molecular sieves. As an example of their use, five isozymes of lactate dehydrogenase have been separated by chromatography on DEAE-Sephadex.

1.6. Partition methods. The technique of counter-current distribution between immiscible solvents (Craig, 1960*b*, 1962) is similar in principle to column chromatography, but it has the advantage that adsorption with the resultant 'tailing' of bands is eliminated. Solutes with quite similar partition coefficients can be separated by multiple distributions between two liquid phases. The two phases are previously equilibrated so that no volume change occurs during the separation. Equal volumes of one phase, usually the lower for convenience, are placed in a series of tubes and the solute and upper phase are introduced into the first tube. After equilibration has been induced by mechanical agitation, the two phases are allowed to separate and the upper phase is transferred to fresh lower phase in the second tube. New upper phase is introduced into the first tube. Both tubes are now equilibrated and the upper phases are transferred, that from the second tube into the third tube containing fresh lower phase and that from the first into the second tube. The process is continued after the introduction of fresh upper phase into the first tube. The number of operations increases rapidly with the number of tubes in use and manual operation is exceedingly tedious for more than a few transfers. Craig, however, has designed automatic machines which mechanically agitate the tubes, allow the phases to separate and transfer all the upper phases simultaneously. Using this type of equipment, it is possible to carry out hundreds of transfers. Under these conditions, the concentration of a solute plotted against numbers of tubes approximates to a Gaussian distribution comparable to the distribution of solute through a peak in column chromatography when behaviour approximates to ideal. It is clear that the stationary lower phase in counter-current distribution may be compared to the stationary phase in column partition chromatography. A mixture of solutes will, if the partition coefficients are sufficiently different, form a series of peaks. Even if two substances cannot be separated, the distribution curve is likely to be skewed so that departure from Gaussian behaviour can be detected. To this extent, the technique is useful also as a criterion of purity.

Craig has used counter-current distribution extremely effectively for the isolation of individual members of a family of related peptide antibiotics such as the tyrocidines and bacitracins. During determinations of amino-acid sequences of these antibiotics, peptides obtained by partial acid hydrolysis were also separated by counter-current distribution. On the other hand, partition methods have not found wide application for the separation of proteins; the choice of solvents is limited, especially if denaturation is to be avoided.

1.7. Electrophoresis. Proteins contain groups which have widely different pK_a values (see p. 57). At a low pH, amino groups are protonated and carboxyl groups tend not to be anionic; hence, the molecule carries a net positive charge. Conversely, at a high pH, carboxyl groups are dissociated and amino groups tend not to be protonated; hence, the molecule carries a net negative charge. At the isoelectric point, the molecule has an equal number of positively and negatively charged groups and no net charge.

Like simple inorganic ions, proteins migrate in an electric field. Below the isoelectric point, they are cationic and migrate to the cathode, while above the isoelectric point, they are anionic and migrate to the anode. At the isoelectric point, no migration is expected. This simple concept is not strictly correct in practice owing to the phenomenon of electro-osmosis. The mobility of a protein in an electric field is proportional to the net charge on the molecule and to its diffusion coefficient, which in turn depends on the molecular weight and shape (see p. 45). It is possible, therefore, to separate proteins by electrophoresis provided that they differ sufficiently in the state of charge of the molecule, molecular weight or shape. Since the first can be controlled by adjustment of pH, electrophoresis is a versatile technique for protein separation. There are different experimental methods; some can be used preparatively while others are more suitable as criteria of purity. Broadly, there are two chief electrophoretic techniques; boundary and zone electrophoresis.

In boundary electrophoresis, the protein is allowed to migrate in free solution. The temperature must be rigorously controlled to prevent thermal diffusion and movement of the protein is followed by an optical system which makes use of the change of refractive index with concentration. Migration must occur in a vertical tube and the solution below any boundary must be denser than above it, since gravitational convection would upset the system. Typically, electrophoresis is carried out in a U-tube. In the descending side of the U-tube, the slowest component will form the uppermost boundary, while in the ascending side of the U-tube, the fastest component will form the uppermost boundary. Other boundaries may be observed which represent mixtures of these and other components. The equipment and experimental techniques are complicated and are described in a textbook edited by Bier (1959). Boundary electrophoresis is chiefly used as a criterion of purity especially since migration patterns can be studied over a range of pH values and ionic strengths. Since the mobility depends on molecular weight, boundary electrophoresis can be used for determination of molecular weight.

Zone electrophoresis is the other main method for protein separation in an electric field. When an inert support is used for electrophoresis, the components of a mixture form discrete zones, thermal and gravitational diffusion is minimized or eliminated and, in consequence, experimental techniques are greatly simplified. Moreover, zone electrophoresis can be used on a microscale as a criterion of purity or on a larger scale for preparative purposes. Finally, in some experimental methods where adequate cooling can be provided, high voltages can be used to obtain rapid separations of mixtures. Sheets of filter paper and cellulose acetate have been used widely for small scale work. Supports of partly hydrolysed starch-gel, silica-gel or polyacrylamide gel in troughs give excellent resolutions of proteins and it is probable that migration of the components of a mixture depends on molecular size and shape as well as the state of electric charge. Larger scale electrophoresis can be carried out using columns of cellulose or starch. Alternatively, a solution of a mixture of proteins can be applied at a constant rate to a hanging sheet of paper; buffer is allowed to flow down the sheet as in ordinary paper chromatography and the potential is applied across the paper. The components of the mixture are collected in tubes at multiple take-off points at the bottom of the sheet. One of the most important applications of microscale electrophoresis of proteins is found in the separation and quantitative determination, after staining with suitable dyestuffs, of serum proteins in various pathological conditions. More detailed accounts of the theory and practice of zone electrophoresis are available (Morris & Morris, 1963).

1.8. Criteria of purity. Crystallinity is an unreliable indication of the purity of a protein, since a number of crystalline proteins have been shown to consist of mixtures. In addition, proteins do not have melting-points. As mentioned above (§**1.6**), adherence to a Gaussian distribution during counter-current distribution is a useful but rather laborious criterion of purity. Another useful technique is to determine the amount of protein dissolved by a constant volume of solvent at constant temperature when the solvent is equilibrated with different amounts of protein. If the protein is homogeneous, a graph of the amount of protein in solution versus total protein present will consist of two linear portions, one of which has zero slope (fig. 1.7). If the protein is heterogeneous, the graph may be nonlinear at low concentrations of protein and have a finite slope at high concentrations.

Apart from these tests, the best indication of homogeneity is the consistent failure to detect heterogeneity when several analytical techniques such as column chromatography, counter-current distribution and electrophoresis are applied. In addition, ultracentrifugation (p. 44) is a valuable method of detecting impurities. Clearly, when a criterion of purity depends on a negative result in

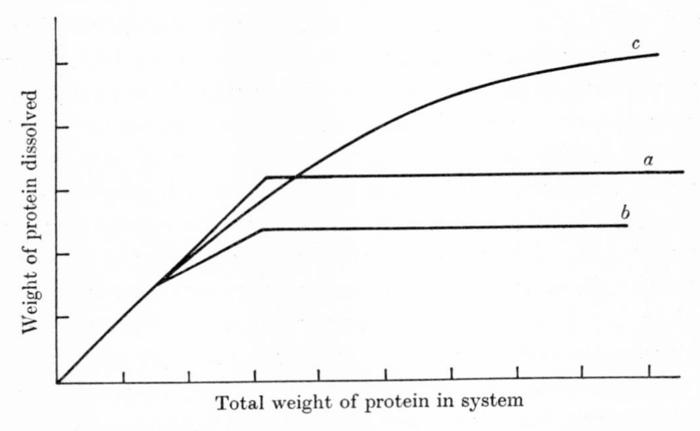

Fig. 1.7. Solubility criterion of purity for proteins: (*a*) homogeneous protein gives two linear curves of unit and zero slope; (*b*) mixture of two proteins which do not form a solid solution; (*c*) mixture of proteins which form a solid solution.

testing for impurities, the confidence which is attached to a claim for homogeneity in a protein preparation is proportional to the number of techniques employed. It should be remembered that each technique can and should be used under different conditions of pH and ionic strength.

Many proteins tend to associate in solution to form dimers, trimers and higher polymers. For a long time, the molecule of insulin was believed to be twice as big as it actually is because it dimerizes readily. Extreme examples of this type of behaviour are found with viruses. For example, tobacco mosaic virus has a molecular weight of approximately 4×10^7 and consists of 5 per cent RNA and 95 per cent protein. The latter consists of more than 2000 identical subunits having a molecular weight of about 17,000. Molecular association can give rise to a misleading indication of heterogeneity. Recognition of this type of behaviour can often be achieved by applying analytical techniques at different ionic strengths or different concentrations of protein. Analysis of terminal amino acids (§**2.5** and §**2.6**) is another useful way of detecting not only molecular association but also heterogeneity, since there is a reasonable chance that two proteins will not have the same amino acids as terminal residues.

2. Determination of the primary structure of proteins

2.1. Introduction. Proteins are built up from α-amino acids, which normally have the L-configuration (I). The side-chain, R, may be alkyl, aromatic or heterocyclic and may contain other functions such as amino, hydroxyl, guanidino or carboxyl groups (table 2.1). The imino acid, proline (II), and cystine (III), which arises by oxidation of two cysteinyl residues, have structures which do not conform to the pattern of the amino acids in the table. It should be remembered that amino acids normally exist as dipolar ions in the crystalline state and in neutral, aqueous solution.

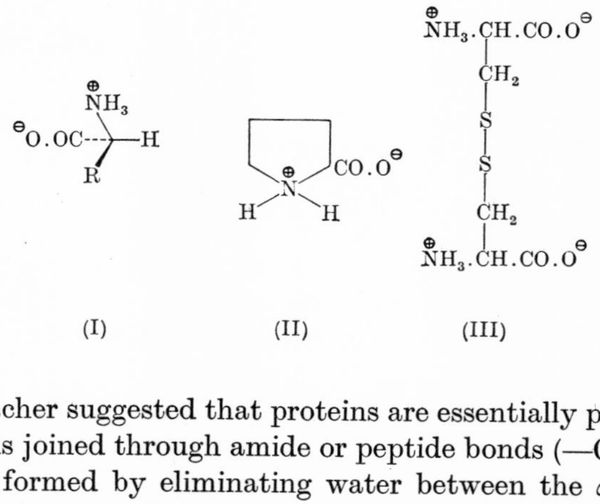

$$\overset{\ominus}{O}.OC\text{---}C(\overset{\oplus}{N}H_3)(R)\text{---}H$$

(I)

$$\text{(pyrrolidine ring)}\ \overset{\oplus}{N}H_2 \text{ with } CO.\overset{\ominus}{O}$$

(II)

$$\begin{array}{l} \overset{\oplus}{N}H_3.CH.CO.\overset{\ominus}{O} \\ \qquad | \\ \quad\ CH_2 \\ \qquad | \\ \quad\ \ S \\ \qquad | \\ \quad\ \ S \\ \qquad | \\ \quad\ CH_2 \\ \qquad | \\ \overset{\oplus}{N}H_3.CH.CO.\overset{\ominus}{O} \end{array}$$

(III)

Emil Fischer suggested that proteins are essentially polymers of amino acids joined through amide or peptide bonds (—CO.NH—) which are formed by eliminating water between the α-carboxyl and α-amino groups of adjacent residues. This hypothesis has been supported by overwhelming evidence. For example, proteins contain relatively few titratable amino and carboxyl groups compared with the number of amino acids present. Amino and carboxyl groups appear simultaneously, however, when proteins are degraded by proteolytic enzymes, acid or alkali. Secondly, peptides

TABLE 2.1. *α-Amino acids which occur in proteins*

Amino acid	IUPAC abbreviation	R
Glycine	Gly	—H
Alanine	Ala	$—CH_3$
Valine	Val	$—CH.(CH_3)_2$
Leucine	Leu	$—CH_2.CH(CH_3)_2$
Isoleucine	Ile	$—CH(CH_3).CH_2.CH_3$
Aspartic acid	Asp	$—CH_2.CO.OH$
Asparagine	$Asp(NH_2)$	$—CH_2.CO.NH_2$
Glutamic acid	Glu	$—CH_2.CH_2.CO.OH$
Glutamine	$Glu(NH_2)$	$—CH_2.CH_2.CO.NH_2$
Serine	Ser	$—CH_2.OH$
Threonine	Thr	$—CH(CH_3).OH$
Cysteine	CySH	$—CH_2.SH$
Methionine	Met	$—CH_2.CH_2.S.CH_3$
Lysine	Lys	$—CH_2.CH_2.CH_2.CH_2.NH_2$
Arginine	Arg	$—CH_2.CH_2.CH_2.NH.C(:NH).NH_2$
Phenylalanine	Phe	$—CH_2$—(phenyl ring)
Tyrosine	Tyr	$—CH_2$—(ring)—OH
Tryptophan	Trp	$—CH_2$—(indole ring, N H)
Histidine	His	$—CH_2$—(imidazole ring, N, NH)
Proline	Pro	See II
Cystine	CyS CyS (bracketed)	See III

have been synthesized which are identical to fragments obtained by partial hydrolysis of proteins. Some of the smaller proteins have been chemically synthesized by unambiguous routes and shown to possess the full biological activity of the natural proteins. Thirdly, proteolytic enzymes hydrolyse proteins and synthetic peptides and display the same specificity for particular amino-acid residues in both types of substrate.

Other types of linkages are not excluded; there is conclusive evidence for the presence of disulphide bonds which can bridge two cysteine residues in the same or different polypeptide chains. The two possibilities respectively give rise to loops within a chain and cross-linkages between chains. Cross-linkage may also occur through phospho-diester linkages involving the hydroxyl groups in the side chains of serine (IV) and threonine.

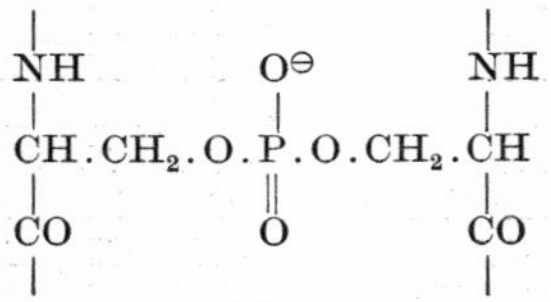

$$\begin{array}{ccc} | & & | \\ NH & O^{\ominus} & NH \\ | & | & | \\ CH.CH_2.O.&P&.O.CH_2.CH \\ | & \| & | \\ CO & O & CO \\ | & & | \end{array}$$

(IV)

Several unusual types of linkages have been suggested to explain the chemical and physical properties of elastin and collagen (Partridge, 1962; Harding, 1965). Ester linkages, involving the β-carboxyl group of aspartic acid and the γ-carboxyl group of glutamic acid on the one hand and aliphatic hydroxyl groups of serine and hydroxyproline on the other, may be present. The ϵ-amino groups of lysine may also be involved in cross-linking. In the case of elastin, there is stronger evidence that two unusual amino acids, desmosine (V) and isodesmosine (VI), which are biosynthesized from lysine, are sites for cross-linking.

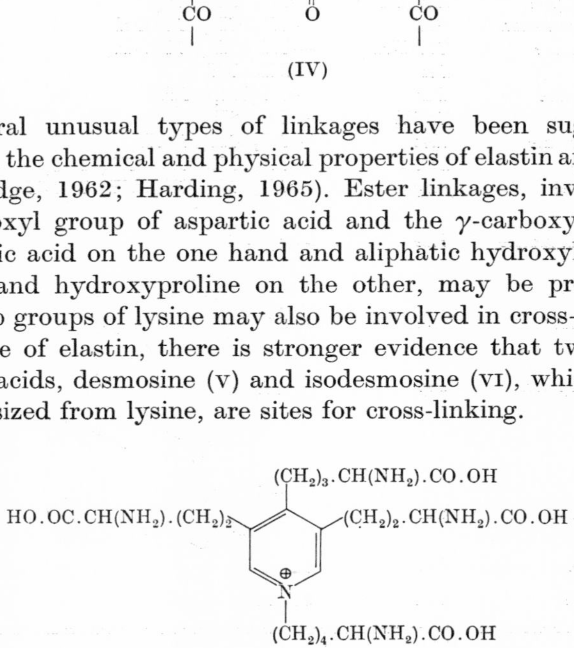

$(CH_2)_3.CH(NH_2).CO.OH$

$HO.OC.CH(NH_2).(CH_2)_2$ — $(CH_2)_2.CH(NH_2).CO.OH$

$N^{\oplus}$

$(CH_2)_4.CH(NH_2).CO.OH$

(V)

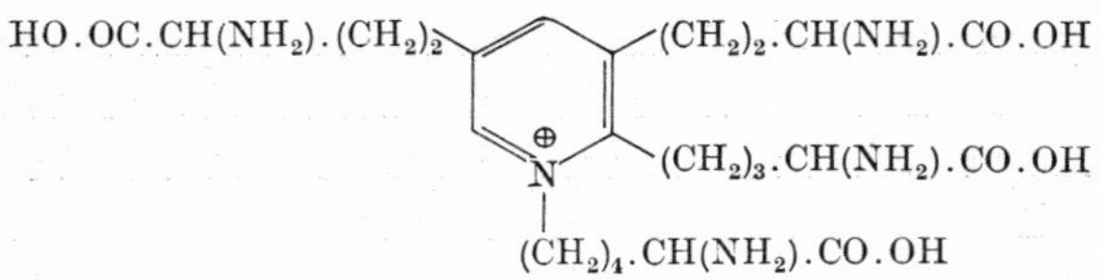

$HO.OC.CH(NH_2).(CH_2)_2$ — $(CH_2)_2.CH(NH_2).CO.OH$

$N^{\oplus}$ — $(CH_2)_3.CH(NH_2).CO.OH$

$(CH_2)_4.CH(NH_2).CO.OH$

(VI)

2.2. General strategy. The structure of a protein can be defined in terms of its amino-acid sequence (primary structure), the conformation of the polypeptide chain resulting from regular (e.g. α-helix) or irregular folding (secondary and tertiary structures), and the association of more than one chain to form a stable unit (quaternary structure). The primary structure is considered in this chapter; secondary, tertiary and quaternary structures are dealt with in chapter 3.

In order to determine the primary structure of a protein it is necessary (*a*) to conduct a qualitative and quantitative analysis of the amino acids present; (*b*) to determine the molecular weight by suitable physical methods in order to calculate the number of residues of each amino acid present (see chapter 3); (*c*) to determine the number of polypeptide chains present in the molecule by quantitative analysis of *N*- and *C*-terminal amino-acid residues (an *N*-terminal residue has a free α-amino group while a *C*-terminal residue has a free α-carboxyl group); (*d*) to separate the chains if more than one is present in the molecule; (*e*) to cleave each chain, preferably by specific methods, into fragments whose sequence can be determined by, for example, partial hydrolysis or by stepwise cleavage and identification of one amino acid at a time from one terminus of the chain. Complications can arise under step (*c*); the *N*-terminal amino acid may be *N*-acylated or the *C*-terminal amino acid may be present as its amide. In either case, the number of *N*- and *C*-termini will be unequal. In the event that both *N*- and *C*-termini are substituted or that the polypeptide is cyclic, no terminal groups will be detectable. Frequently, proteins contain covalently bound groups, known as prosthetic groups, which are not composed of amino acids. Examples of prosthetic groups include haem in haemoglobins, cytochrome *c* and peroxidases, biotin in carboxylases and carbohydrates in glycoproteins. If a prosthetic group is present, its position and mode of attachment to the polypeptide chain must be determined.

It is worth examining a hypothetical case to appreciate the strategical and logical problems involved in the elucidation of the primary structure of a polypeptide. Let us suppose that a peptide has the following composition:

$$\mathrm{Gly_3,Ala_3,Lys_2,Phe_2,Thr_2}$$

and that alanine is *N*-terminal while phenylalanine is *C*-terminal. We can now write a partial sequence:

$$\mathrm{Ala\text{-}(Gly_3,Ala_2,Lys_2,Phe,Thr_2)\text{-}Phe.}$$

Let us suppose that the polypeptide gives on partial acid hydrolysis the following dipeptides:

Ala-Lys; Lys-Gly; Gly-Phe; Phe-Thr; Thr-Ala; Gly-Thr; Ala-Gly.

Some higher peptides and some free amino acids will inevitably be formed, but we will suppose that conditions can be chosen which lead preferentially to the above dipeptides and that these can be separated by suitable chromatographic procedures. The sequence of amino acids in each dipeptide could be established by identification of the *N*- and/or *C*-terminal residues and by complete hydrolysis followed by identification of the constituent amino acids. If it is assumed that the seven dipeptides are the only ones which can arise from the original polypeptide, a number of sequences are possible (VII–XII):

Ala-Lys-Gly-Phe-Thr-Ala-Lys-Gly-Thr-Ala-Gly-Phe, (VII)
Ala-Lys-Gly-Thr-Ala-Gly-Phe-Thr-Ala-Lys-Gly-Phe, (VIII)
Ala-Lys-Gly-Thr-Ala-Lys-Gly-Phe-Thr-Ala-Gly-Phe, (IX)
Ala-Gly-Thr-Ala-Lys-Gly-Phe-Thr-Ala-Lys-Gly-Phe, (X)
Ala-Lys-Gly-Phe-Thr-Ala-Gly-Thr-Ala-Lys-Gly-Phe, (XI)
Ala-Gly-Phe-Thr-Ala-Lys-Gly-Thr-Ala-Lys-Gly-Phe. (XII)

If other dipeptides, in addition to the seven given above, were formed but could not be separated from the partial hydrolysate, many other sequences would be possible, e.g.:

Ala-Phe-Thr-Ala-Lys-Gly-Thr-Lys-Gly-Ala-Gly-Phe. (XIII)

It will be seen that sequence (XIII) contains all seven of the dipeptides identified. If this were the correct sequence, however, three other dipeptides, Ala-Phe, Thr-Lys and Gly-Ala, should be formed.

Let us suppose that, in addition to the seven dipeptides, the following nine tripeptides can be isolated from a partial hydrolysate and identified:

Ala-Lys-Gly; Lys-Gly-Phe; Gly-Phe-Thr; Phe-Thr-Ala; Thr-Ala-Lys;
Lys-Gly-Thr; Gly-Thr-Ala; Thr-Ala-Gly; Ala-Gly-Phe.

The sequences of the tripeptides would follow from determinations of their amino-acid composition and identification of the *N*- and *C*-terminal amino acids. With this information, structures (X), (XI) and (XII) can be excluded; Lys-Gly-Thr, Ala-Gly-Phe and Thr-Ala-Gly could not arise from sequence (X), the first two of these tripeptides could not arise from sequence (XI), while the last of these three tripeptides is incompatible with the sequence (XII). It is still impossible, however, to decide between the sequences (VII–IX). An unequivocal solution would require the isolation and identification of some tetrapeptides from the partial hydrolysis mixture.

Let us now examine the consequences of using methods of fragmenting the original polypeptide which are more specific than partial acid hydrolysis. The proteolytic enzymes, trypsin and chymotrypsin, are suitable agents for achieving this. The specificities of these enzymes are discussed later (§**2.8**); it is sufficient to state here that fragments obtained by trypsin-catalysed hydrolysis will have *C*-terminal lysine, while chymotrypsin will afford peptides with *C*-terminal phenylalanine. Since the original polypeptide has two lysyl residues, neither of which is *C*-terminal, it would be expected that hydrolysis with trypsin would give three fragments. Let us suppose that these can be separated and that determination of their amino-acid composition and *N*-terminal residues leads to the following partial structures for the three fragments:

Ala-Lys; Gly-(Phe,Thr,Ala)-Lys; Gly-(Thr,Ala,Gly)-Phe.

Since the original polypeptide has *N*-terminal alanine and *C*-terminal phenylalanine, the order of the fragments in the original molecule is known and we can write the following partial structure:

Ala-Lys-Gly-(Phe,Thr,Ala)-Lys-Gly-(Thr,Ala,Gly)-Phe.

It will be noted that structures (VIII–XII) can already be eliminated.

Since phenylalanine is *C*-terminal in the original polypeptide, only two fragments would be expected from the hydrolysis with chymotrypsin. Let us suppose that they have the following partial sequences:

Ala-(Lys,Gly)-Phe; Thr-(Ala_2,Gly_2,Thr,Lys)-Phe.

When this information is combined with that derived from the tryptic hydrolysis, the following partial sequence is obtained:

Ala-Lys-Gly-Phe-Thr-Ala-Lys-Gly-(Thr,Ala,Gly)-Phe.

If the peptide, Gly-(Thr,Ala,Gly)-Phe, arising from the trypsin-catalysed hydrolysis of the polypeptide, were subjected to partial acid hydrolysis, identification of any two dipeptides would provide the minimum amount of information required to complete the sequence (VII). It should be stressed, however, that it is essential to have confirmatory evidence and the peptides derived from enzymic hydrolyses would normally be subjected to partial acid hydrolysis or stepwise degradation (see p. 29) through several steps. Alternatively, or in addition, peptides obtained by hydrolysis with one enzyme would be further degraded with another enzyme. For example, the peptide, Gly-(Phe,Thr,Ala)-Lys, obtained by hydrolysis with trypsin, would give Gly-Phe and Thr-Ala-Lys when hydrolysed with chymotrypsin.

The example given is a simple case and the sequence could probably be derived by stepwise degradation alone rather than by the procedure outlined above. Since, however, partial hydrolysis provides inconclusive results with this dodecapeptide, it is obvious that it would be even less efficient with large molecules. It should now be clear that the best approach is to cleave the protein into as few fragments as possible, to separate these and to cleave each again by another method. When suitably small fragments are obtained, stepwise degradation or partial hydrolysis will lead to a complete sequence for the fragment.

2.3. Amino-acid analysis. The rapid progress in the elucidation of the primary structure of proteins, which has occurred in recent years, would not have been possible but for the development of methods for the separation and determination of amino acids on the microscale. For the determination of most amino acids, the protein must first be completely hydrolysed. The commonest method involves hydrolysis with 6 N-hydrochloric acid in a sealed, evacuated tube at 110 °C for twenty-four to seventy-two hours, although some difficulties are encountered. During hydrolysis, black humin may form especially if the protein contains carbohydrate. Humin formation is minimized by carrying out the hydrolysis *in vacuo* in a relatively large volume of hydrochloric acid which is free of heavy metal ions. Tryptophan is largely destroyed by acid hydrolysis. Asparagine and glutamine are completely converted into aspartic and glutamic acids respectively. Cysteine and cystine are partially destroyed during acid hydrolysis and better results are obtained by previously oxidizing them with performic acid to cysteic acid which is stable. Serine and threonine are slowly destroyed during hydrolysis and it is customary to carry out analyses on samples of protein which have been hydrolysed for different periods of time. The content of serine and threonine can then be extrapolated back to zero time. Finally, valine and isoleucine are only slowly liberated by acid hydrolysis, presumably because their bulky side chains cause steric hindrance. For these amino acids, analytical values are extrapolated to infinite time.

Alkaline hydrolysis destroys arginine, cysteine, cystine, serine and threonine and is therefore useless for the complete analysis of the amino acids in a protein. Tryptophan, however, largely survives alkaline hydrolysis and so the method has a limited use for the analysis of this amino acid. The protein is usually hydrolysed with 5 N-barium hydroxide under reflux for eighteen to twenty-four hours.

Hydrolysis of proteins with proteolytic enzymes is attractive in principle, since amino acids such as tryptophan, asparagine, glutamine, serine and threonine are not destroyed. It is difficult, however, to achieve complete hydrolysis and, moreover, autolysis of the enzyme may contribute amino acids to the hydrolysate from the protein. 'Pronase', from *Streptomyces griseus*, is probably the least specific proteolytic enzyme known and hydrolyses proteins to the extent of 60 to 90 per cent. More satisfactory results have been obtained by using papain followed by leucine aminopeptidase and prolidase. Amino-acid analyses of the hydrolysates of several proteins obtained in this way have given results which agree with those from acid hydrolysis.

Ion-exchange chromatography is the most popular method for the quantitative analysis of a mixture of amino acids. There are several technical variations, but basically the mixture of amino acids, derived from about 3 mg. of protein, is resolved on a column of a sulphonated polystyrene resin by elution with buffers (fig. 2.1). In the earlier methods, the effluent was fractionally collected and manually analysed, by a colorimetric procedure, using the ninhydrin reagent. More recently, the process has been mechanized; buffer is pumped through the column at a fixed rate, the effluent is mixed with a controlled flow of ninhydrin and heated by passage through a reaction coil at 100°. The developed colour is measured in a flow-colorimeter and the extinction is plotted on a recorder and integrated through a peak. The results have a precision of 100 ± 3 per cent. The amino acids can be identified from their elution positions. Further details can be found in an article by Light & Smith (1963).

There are other methods of analysis which involve conversion of the amino acids into various derivatives. Although demanding an extra step with concomitant losses, some of these techniques have potential advantages. In one method, the amino acids are converted into yellow 2,4-dinitrophenyl-(DNP-) derivatives (§**2.5**).

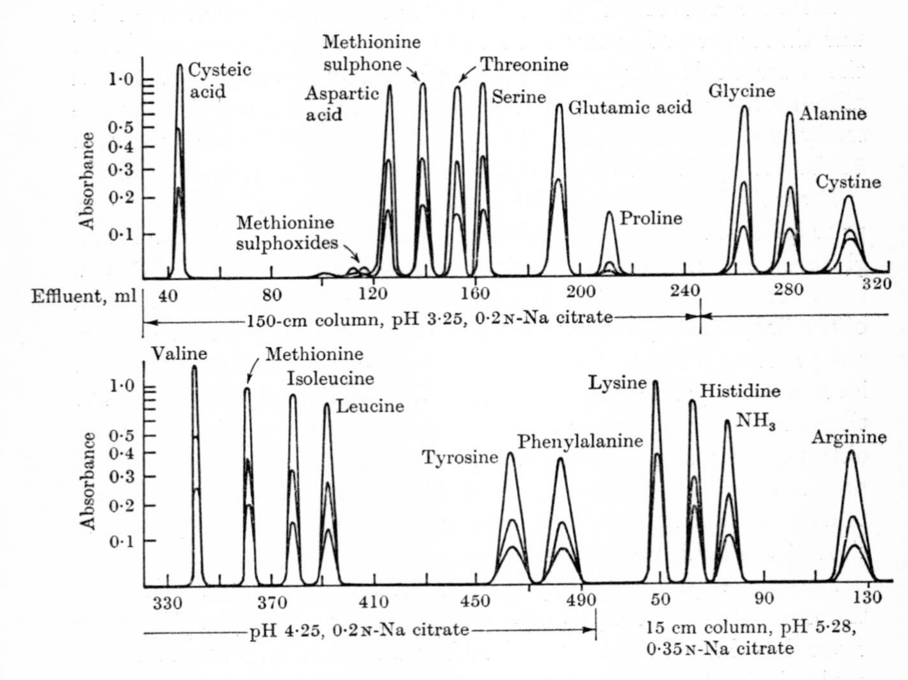

Fig. 2.1. Ion-exchange chromatography of amino acids on a column of sulphonated polystyrene resin (Amberlite IR-120). The optical density of the products formed by reaction with ninhydrin is automatically recorded at three wavelengths. [D. H. Spackman, W. H. Stein & S. Moore (1958). *Analyt. Chem.* **30**, 1190.]

These can be separated by two-dimensional paper chromatography, eluted from the paper and determined spectrophotometrically. Correction factors are used to compensate for slightly different recoveries for each DNP-amino acid. About 0·1 μmole of each derivative is required and the results have a precision of 100 ± 4 per cent.

Gas-liquid chromatography, because of its sensitivity and speed of separation, has also been studied. It is necessary to convert amino acids into volatile compounds such as the methyl esters of the *N*-trifluoroacetyl derivatives (Karmen & Saroff, 1964).

In contrast to the methods so far described, analyses for individual amino acids can be carried out on the original protein hydrolysate without the need for chromatography by using specific enzymes such as amino acid decarboxylases. Carbon dioxide, which is evolved, is determined manometrically:

$$R.CH(NH_2).CO.OH \rightarrow R.CH_2.NH_2 + CO_2.$$

Several specific decarboxylases have been isolated from bacterial sources. A further useful application of enzymic assays allows the configuration of the amino acid to be determined. Although amino acids in proteins normally have the L-configuration, D-amino acids occur in several peptide antibiotics. A D-amino acid oxidase from kidney and an L-amino acid oxidase from snake venoms catalyse the following reaction with amino acids of the appropriate configuration:

$$R.CH(NH_2).CO.OH + O_2 + H_2O \rightarrow R.CO.CO.OH + NH_3 + H_2O_2.$$

The reaction can be followed by measuring manometrically the oxygen used or by determining colorimetrically the ammonia evolved.

Some amino acids can be determined without hydrolysing the protein. For example, several methods are available for the determination of the thiol group of cysteine (Benesch & Benesch, 1962). Some of these depend on the formation of mercaptides with heavy metals and especially with organic mercury compounds:

$$R'.Hg.X + R.SH \rightarrow R'.Hg.S.R + HX.$$

The reaction may be followed amperometrically or polarographically, or sodium nitroprusside may be used as an indicator. Thiols react with *N*-ethylmaleimide (XIV) rapidly and quantitatively:

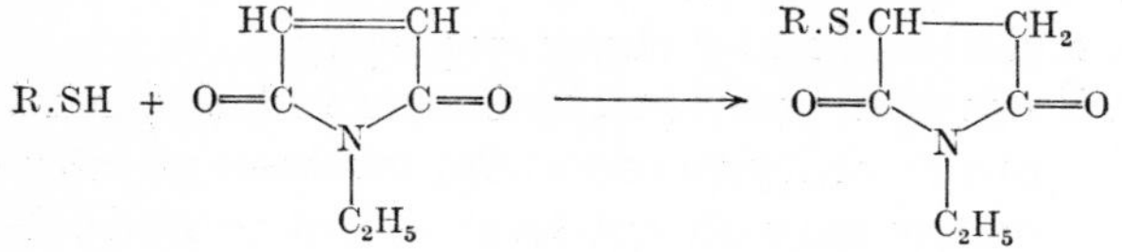

```
             HC====CH                   R.S.CH----CH2
             |     |                        |     |
R.SH  +  O==C      C==O   ---------->   O==C      C==O
             \    /                          \   /
               N                               N
               |                               |
              C2H5                            C2H5

             (XIV)                           (XV)
```

The reagent, but not the product (XV), absorbs light at 300 mμ; the course of the reaction can thus be followed spectrophotometrically.

2.4. Cleavage of disulphide bridges. When a protein contains cystine, irrespective of whether this forms a loop within a chain or links two chains together, it is usual, except when the position of the disulphide bridge is under investigation, to cleave the disulphide bridge before carrying out sequence studies. This can be achieved by oxidation with performic acid, each cystinyl residue giving rise to two cysteic acid residues:

$$\begin{array}{lcl}
\begin{array}{l} | \qquad\qquad\qquad\qquad\quad\; | \\ \mathrm{NH} \qquad\qquad\qquad\quad \mathrm{NH} \\ | \qquad\qquad\qquad\qquad\quad\; | \\ \mathrm{CH.CH_2.S.S.CH_2.CH} \\ | \qquad\qquad\qquad\qquad\quad\; | \\ \mathrm{CO} \qquad\qquad\qquad\quad\;\; \mathrm{CO} \\ | \qquad\qquad\qquad\qquad\quad\; | \end{array}
& \xrightarrow{\mathrm{H.CO.O.OH}} &
\begin{array}{l} | \qquad\qquad\qquad\qquad\qquad\qquad\qquad\quad | \\ \mathrm{NH} \qquad\qquad\qquad\qquad\qquad\qquad\quad \mathrm{NH} \\ | \qquad\qquad\qquad\qquad\qquad\qquad\qquad\quad | \\ \mathrm{CH.CH_2.SO_3H + HO_3S.CH_2.CH} \\ | \qquad\qquad\qquad\qquad\qquad\qquad\qquad\quad | \\ \mathrm{CO} \qquad\qquad\qquad\qquad\qquad\qquad\quad\;\; \mathrm{CO} \\ | \qquad\qquad\qquad\qquad\qquad\qquad\qquad\quad | \end{array}
\end{array}$$

Alternatively, the disulphide bridges can be reductively cleaved. Several methods have been used, but the reagent of choice appears to be mercaptoethanol. When reduction is complete, the liberated thiol groups in the protein are alkylated with iodoacetic acid; cysteine is thus converted into *S*-carboxymethylcysteine. Although mercaptoethanol reacts with iodoacetic acid, it does so more slowly than cysteine. The competitive carboxymethylation of mercaptoethanol with iodoacetic acid is sufficiently rapid, however, to prevent appreciable formation of sulphonium derivatives by the reaction of iodoacetic acid with the side-chains of methionyl residues.

2.5. Identification of *N*-terminal groups. There are two main types of method for identifying the *N*-terminal residue of a peptide or protein. In the first, a suitable group is attached to the α-amino group of the *N*-terminal residue so that, after hydrolysis with acid or enzymes, the labelled residue can be sensitively detected by spectrophotometry, fluorescence or radioactive counting and identified by chromatography.

The DNP-method, developed by Sanger and used in his classical elucidation of the structure of insulin, has been mentioned above (p. 24) in connection with the assay of amino acids. The DNP-group is introduced by reaction with 1-fluoro-2,4-dinitrobenzene (XVI) in a buffer at about pH 8. Reaction is likely to occur also with the ϵ-amino groups of lysine, the phenolic hydroxyl group of tyrosine, the thiol group of cysteine and the imidazole nucleus of histidine. Complete hydrolysis of the DNP-peptide (XVII) with acid gives the yellow DNP-derivative (XVIII) of the *N*-terminal residue, together with ε-DNP-lysine, *O*-DNP-tyrosine and free amino acids, which can be identified by chromatography. Thin-layer chromatography is particularly useful for rapid, qualitative work. Unfortunately, some DNP-derivatives such as DNP-proline and DNP-glycine are rather unstable to acid.

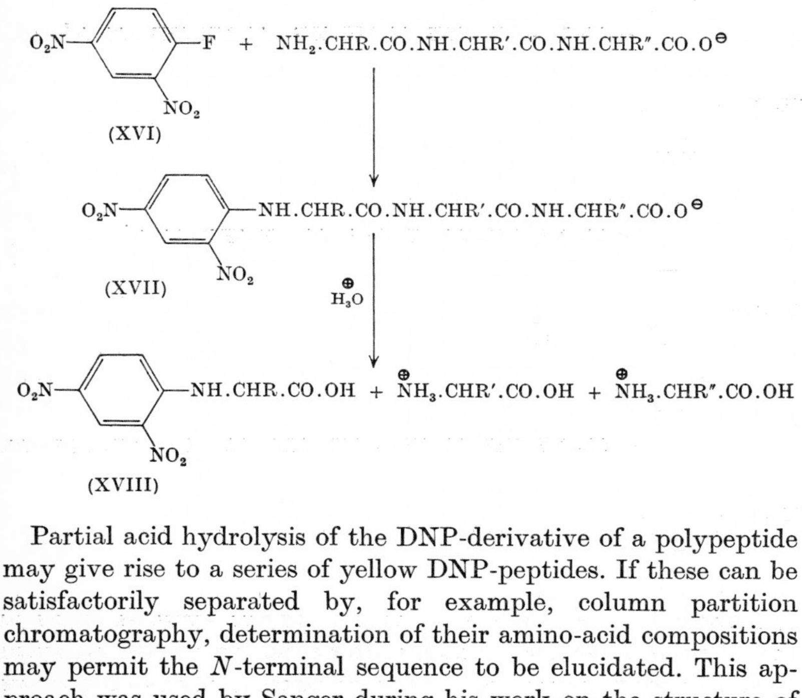

O_2N–C$_6$H$_3$(NO$_2$)–F (XVI) + NH_2.CHR.CO.NH.CHR′.CO.NH.CHR″.CO.$O^{\ominus}$

↓

O_2N–C$_6$H$_3$(NO$_2$)–NH.CHR.CO.NH.CHR′.CO.NH.CHR″.CO.$O^{\ominus}$ (XVII)

↓ $\overset{\oplus}{H_3O}$

O_2N–C$_6$H$_3$(NO$_2$)–NH.CHR.CO.OH (XVIII) + $\overset{\oplus}{N}H_3$.CHR′.CO.OH + $\overset{\oplus}{N}H_3$.CHR″.CO.OH

Partial acid hydrolysis of the DNP-derivative of a polypeptide may give rise to a series of yellow DNP-peptides. If these can be satisfactorily separated by, for example, column partition chromatography, determination of their amino-acid compositions may permit the *N*-terminal sequence to be elucidated. This approach was used by Sanger during his work on the structure of insulin. Partial hydrolysis of the DNP-derivative of the oxidized *A* chain gave DNP-Gly, DNP-Gly-Ile, DNP-Gly-(Ile,Val) and DNP-Gly-(Ile,Val,Glu). Obviously, the *N*-terminal sequence of the *A* chain of insulin is: Gly-Ile-Val-Glu-.... Similarly, peptides containing ε-DNP-lysine may be isolated from partial acid hydrolysates of the DNP-derivatives of proteins. Since the chromophoric group is in the side-chain in this case, the *N*-terminal residue of the DNP-peptide must be identified by further dinitrophenylation and hydrolysis.

Recently, 1-dimethylaminonaphthalene-5-sulphonyl chloride ('dansyl' chloride) (XIX) has been suggested as an alternative to 1-fluoro-2,4-dinitrobenzene, since the *N*-(1-dimethylaminonaphthalene-5-sulphonyl) amino acids (XX) fluoresce strongly in ultraviolet light and 10^{-10}–10^{-9} moles can be detected. The method is thus about 100 times more sensitive than the DNP-technique. Moreover, 'dansyl'-derivatives of amino acids are stable to acid hydrolysis.

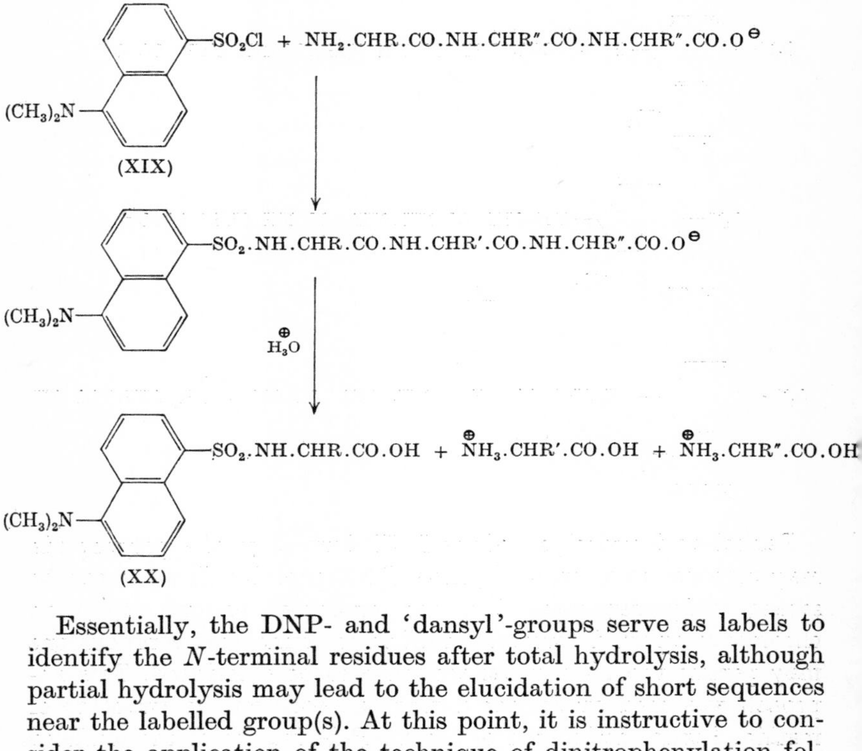

$(CH_3)_2N$–$C_{10}H_6$–SO_2Cl + $NH_2.CHR.CO.NH.CHR''.CO.NH.CHR''.CO.O^{\ominus}$

(XIX)

↓

$(CH_3)_2N$–$C_{10}H_6$–$SO_2.NH.CHR.CO.NH.CHR'.CO.NH.CHR''.CO.O^{\ominus}$

↓ $H_3\overset{\oplus}{O}$

$(CH_3)_2N$–$C_{10}H_6$–$SO_2.NH.CHR.CO.OH$ + $\overset{\oplus}{N}H_3.CHR'.CO.OH$ + $\overset{\oplus}{N}H_3.CHR''.CO.OH$

(XX)

Essentially, the DNP- and 'dansyl'-groups serve as labels to identify the *N*-terminal residues after total hydrolysis, although partial hydrolysis may lead to the elucidation of short sequences near the labelled group(s). At this point, it is instructive to consider the application of the technique of dinitrophenylation followed by partial hydrolysis to the elucidation of the sequence of peptide (VII). Three DNP-groups would be introduced and, from a partial hydrolysate, derivatives such as:

DNP-Ala-Lys(ε-DNP) Lys(α-DNP)-Gly...

DNP-Ala-Lys (DNP on Lys)	Lys-Gly (DNP on Lys)	Ala-Lys (DNP on Lys)	DNP-Ala-Lys-Gly (DNP on Lys)
Lys-Gly-Phe (DNP on Lys)	Thr-Ala-Lys (DNP on Lys)	Ala-Lys-Gly (DNP on Lys)	Lys-Gly-Thr (DNP on Lys)

might well be isolated. Those peptides containing a free α-amino group would be further dinitrophenylated and hydrolysis of all the derivatives would give enough information for their sequences to be elucidated. Unless tetrapeptide derivatives bearing a DNP-group were isolated, however, it would still be impossible to distinguish between three structures (VII–IX).

The second and more powerful type of method for identifying the *N*-terminal residue of a polypeptide involves the attachment of a suitable substituent on the α-amino group followed by the selective cleavage of the modified *N*-terminal residue. Repetition of such a method then provides a means for the stepwise degradation of a peptide and the progressive elucidation of its sequence. Many such procedures have been described in the literature (Thompson, 1960). The best known is due to Edman and involves the reaction between phenyl isothiocyanate and the peptide at approximately pH 8. Reaction occurs with the unprotonated α-amino group and the solution is either buffered or kept at constant pH by the addition of base from a pH-stat. The resultant *N*-phenylthiocarbamoyl-derivative (XXI) is freed from excess phenyl isothiocyanate by extraction with an organic solvent. Under suitable conditions of acid catalysis, it gives rise to the 3-phenyl-2-thiohydantoin (XXII) corresponding to the *N*-terminal residue and the peptide (XXIII) containing one residue less than the original:

$$\overset{\oplus}{N}H_3.CHR.CO.NH.CHR'.CO.NH.CHR''.CO.O^{\ominus}$$

$$\Updownarrow$$

$$C_6H_5.N\!:\!\overset{\delta+}{C}\!:\!\overset{\delta-}{S} + \ddot{N}H_2.CHR.CO.NH.CHR'.CO.NH.CHR''.CO.O^{\ominus}$$

$$\downarrow$$

$$\underset{\text{(XXI)}}{C_6H_5.NH.CS.NH.CHR.CO.NH.CHR'.CO.NH.CHR''.CO.O^{\ominus}}$$

$$\downarrow \overset{\oplus}{H}$$

(XXII): 3-phenyl-2-thiohydantoin ring — S=C, N(H), CHR, C=O, C_6H_5.N

$$\underset{\text{(XXIII)}}{\overset{\oplus}{N}H_3.CHR'.CO.NH.CHR''.CO.OH}$$

Cyclization and degradation (XXI→XXII + XXIII) is more complicated than shown; at least two intermediate steps have been demonstrated. A variety of acid catalysts have been used for this step, but perhaps the best is trifluoroacetic acid. With this reagent, cyclization and degradation proceed rapidly at room temperature with little risk to the residual peptide. Use of dilute hydrochloric acid, sometimes above room temperature, can cause hydrolysis of other peptide bonds so that subsequent steps liberate more than one 3-phenyl-2-thiohydantoin. Most of the 3-phenyl-2-thiohydantoins can be extracted from acid solution with organic solvents; they can be identified by chromatography and determined spectrophotometrically.

Several variations are possible. For example, a quantitative determination of amino acids can be carried out on a small portion of the degraded peptide (XXIII); the *N*-terminal residue, which has been removed as a 3-phenyl-2-thiohydantoin derivative, is identified by subtracting the amino-acid composition from that of the original peptide. Alternatively, the *N*-terminal residue is identified at each cycle of the Edman procedure by subjecting a small portion of the free peptide to 'dansylation' and total hydrolysis followed by chromatography of the 'dansyl'-amino acid. The bulk of the free peptide is used for stepwise degradation. The conventional Edman method has also been adapted so that all the steps can be carried out on a filter-paper support; after each cycle, the 3-phenyl-2-thiohydantoin is eluted from the filter paper and identified. Very recently, Edman & Begg (1967) have described an apparatus which will carry out automatically most of the steps involved in *N*-terminal sequence determinations. The product from the *N*-terminal residue for each cycle is collected separately in a fraction collector and all are identified chromatographically at the end of the stepwise degradation. Fifteen cycles can be completed in twenty-four hours. The yields are about 98 per cent for each cycle and consequently Edman was able to carry the degradation through sixty steps with apomyoglobin. When this apparatus becomes commercially available it will revolutionize the task of determining amino acid sequences.

A recent method of stepwise degradation from the *N*-terminus, which is related to the Edman procedure, involves carbamoylation of the terminal α-amino group with cyanate at pH 7–8. The reaction probably occurs between a cyanic acid molecule and the unprotonated α-amino group. The *N*-carbamoylpeptide (XXIV), on treatment with acid, gives a hydantoin (XXV) corresponding to the *N*-terminal residue and a peptide containing one residue less than the original:

$$\overset{\oplus}{H} + \overset{\ominus}{\overline{N{:}C{:}O}} \qquad \overset{\oplus}{N}H_3.CHR.CO.NH.CHR'.CO.NH.CHR''.CO.O^{\ominus}$$

$$\rightleftharpoons \qquad\qquad \rightleftharpoons$$

$$H.N{:}\overset{\delta+}{C}{:}\overset{\delta-}{O} + \ddot{N}H_2.CHR.CO.NH.CHR'.CO.NH.CHR''.CO.O^{\ominus}$$

$$\downarrow$$

$$NH_2.CO.NH.CHR.CO.NH.CHR'.CO.NH.CHR''.CO.O^{\ominus}$$

(XXIV)

$$\downarrow \overset{\oplus}{H}$$

$$\text{hydantoin ring: } O{=}C\text{—}NH\text{—}CHR\text{—}C({=}O)\text{—}HN\text{—} \text{(ring)} \qquad + \qquad \overset{\oplus}{N}H_3.CHR'.CO.NH.CHR''.CO.OH$$

(XXV)

The hydantoin (XXV), after isolation, can be hydrolysed to the corresponding amino acid which is identified by chromatography.

Finally, aminopeptidases (Smith & Hill, 1960) specifically hydrolyse *N*-terminal amino acids from peptides and proteins. The aminopeptidase from kidney, which has been most thoroughly investigated, preferentially cleaves amino acids with long alkyl side-chains (e.g. leucine); aromatic amino acids are cleaved rather more slowly and peptides terminating in glycine, serine and proline are rather resistant. It is customary to carry out analyses for amino acids on enzymic digests at various intervals of time. With an *N*-terminal sequence such as Leu-Tyr-Gly-···, the sequence can be established with reasonable certainty, since the residues become progressively more difficult to cleave as the reaction proceeds. With the inverse sequence, Gly-Tyr-Leu-···, however, the cleavage of tyrosine would follow as soon as glycine was released and this would be rapidly followed by the liberation of leucine. The sequence would be difficult to elucidate in this case. Another type of aminopeptidase has been discovered recently which is almost specific for the cleavage of *N*-terminal lysine and arginine. A similar dichotomy of specificities is found with carboxypeptidases *A* and *B* (§**2.6**).

2.6. Identification of *C*-terminal groups. Of several chemical methods, which have been investigated for the identification of *C*-terminal groups, only one has found extensive use. Reaction of polypeptides with anhydrous hydrazine at elevated temperatures cleaves the peptide bonds and forms the hydrazide of all amino acids except the *C*-terminal residue:

$$NH_2.\underset{|}{\overset{R}{C}}H.CO.NH.\overset{R'}{C}H.CO.NH.\overset{R''}{C}H.CO.OH \xrightarrow{N_2H_4} NH_2.\overset{R}{C}H.CO.NH.NH_2 + NH_2.\overset{R'}{C}H.CO.NH.NH_2 + NH_2.\overset{R''}{C}H.CO.OH$$

The *C*-terminal amino acid can be separated from the mixture of hydrazides and identified. Unfortunately, recoveries are frequently low; cystine and tryptophan are completely destroyed, while arginine is converted into ornithine and methionine is oxidized to its sulphoxide.

The use of two pancreatic enzymes, carboxypeptidases *A* and *B*, has been much more successful. These enzymes liberate the *C*-terminal residue from a polypeptide and differ in their specificities. Carboxypeptidase *A* has a fairly broad specificity, although it preferentially liberates aromatic amino acids; *C*-terminal lysine, arginine and proline are not hydrolysed. Carboxypeptidase *B*, on the other hand, is much more selective and liberates only the basic amino acids, lysine and arginine when these are *C*-terminal. With both enzymes, as with aminopeptidases, it is customary to carry out analyses for amino acids at intervals on the enzymic digests.

2.7. Selective chemical methods of cleaving peptide bonds. Partial acid hydrolysis has been widely used and, not unnaturally, efforts have been made to find conditions which effect selective cleavage of peptide bonds (Hill, 1965). Hydrolysis of a protein with 0·25 N-acetic acid under reflux preferentially splits the peptide bonds on both sides of aspartyl residues. For example, a peptide, obtained by hydrolysis of human foetal haemoglobin with trypsin, had the composition:

$$Val_2,Asp_2,Glu_2,Ala,Gly_3,Thr,Leu,Arg,(NH_3).$$

Hydrolysis with 0·25 N-acetic acid under reflux gave three peptides (XXVI–XXVIII) with the following sequences:

Val-Asp(NH_2)-Val-Glu (XXVI)
Val-Asp(NH_2)-Val-Glu-Asp (XXVII)
Ala-Gly-Gly-Glu-Thr-Leu-Gly-Arg (XXVIII)

Arginine would be expected to be the *C*-terminal residue of a peptide obtained by trypsin-catalysed hydrolysis (see p. 36) and, in addition, application of the Edman procedure to the original peptide revealed that the first five amino acids corresponded to the peptide (XXVII). Hence, the complete sequence (XXIX) must be:

Val-Asp(NH_2)-Val-Glu-Asp-Ala-Gly-Gly-Glu-Thr-Leu-Gly-Arg (XXIX)

In contrast, peptide bonds involving the amino groups of serine and threonine are preferentially cleaved during hydrolysis of a protein with 10 N-hydrochloric acid at 30 °C. The cleavage is assisted by the neighbouring hydroxyl group in the side-chain of serine or threonine; the peptide rearranges through a hydroxy-oxazolidine tautomer to an *O*-acyl-derivative which is then hydrolysed (XXX–XXXIII).

(XXX) ⇌ (XXXI) $\underset{-\,H^{\oplus}}{\overset{+\,H^{\oplus}}{\rightleftharpoons}}$ (XXXII) ⇌ (XXXIII)

N → *O* acyl migration also occurs in anhydrous acids. The use of concentrated sulphuric acid or formic acid is attended by serious side-reactions such as *O*-formylation, but anhydrous hydrogen fluoride appears to be more satisfactory. After rearrangement has taken place, the liberated amino groups can be formylated and the ester groups are hydrolysed with base. Application of this procedure to a protein (XXXIV → XXXVII), gives a series of peptides with *N*-terminal serine or threonine.

$$\underset{\text{(XXXIV)}}{\text{R.CO.NH.}\overset{\text{HO.CH}_2}{\text{CH}}\text{.CO.NH.R}'} \xrightarrow{\text{HF}} \underset{\text{(XXXV)}}{\overset{\oplus}{\text{N}}\text{H}_3\text{.}\overset{\text{R.CO.O.CH}_2}{\text{CH}}\text{.CO.NH.R}'} \xrightarrow[\text{(CH}_3\text{.CO)}_2\text{O}]{\text{H.CO.OH}}$$

$$\underset{\text{(XXXVI)}}{\text{H.CO.NH.}\overset{\text{R.CO.O.CH}_2}{\text{CH}}\text{.CO.NH.R}'} \xrightarrow{\text{aqueous pyridine}} \text{R.CO.OH} + \underset{\text{(XXXVII)}}{\text{H.CO.NH.}\overset{\text{HO.CH}_2}{\text{CH}}\text{.CO.NH.R}'}$$

Cyanogen bromide cleaves dialkyl sulphides to give an alkyl thiocyanate and an alkyl bromide:

$$\text{R.CH}_2\text{.}\ddot{\text{S}}\text{.CH}_2\text{.R} + \text{N}\vdots\text{C.Br} \longrightarrow \text{R.CH}_2\text{.S.C}\vdots\text{N} + \text{R.CH}_2\text{.Br}$$

This reagent has been used to cleave peptide bonds which contain the carbonyl group of a methionyl residue. The reaction is more complicated than with a simple dialkyl sulphide, since ring-closure to homoserine lactone and fission of the peptide bond occur (XXXVIII → XL):

$$\underset{\text{(XXXVIII)}}{\text{R.CO.NH.CH(CH}_2\text{CH}_2\ddot{\text{S}}\text{—CH}_3\text{)—C(=O)—}\ddot{\text{N}}\text{H.R}'} + \text{N}\equiv\text{C—Br} \longrightarrow \underset{\text{(XXXIX)}}{\text{R.CO.NH.CH—C}(=\overset{\oplus}{\text{N}}\text{H.R}')\text{ (lactone ring: CH}_2\text{—CH}_2\text{—O)}\ \text{Br}^{\ominus}} + \text{CH}_3\text{.S.C}\vdots\text{N} \xrightarrow{\text{H}_2\text{O}} \underset{\text{(XL)}}{\text{R.CO.NH.CH—C=O (lactone ring: CH}_2\text{—CH}_2\text{—O)}} + \text{R}'\text{.}\overset{\oplus}{\text{N}}\text{H}_3\overset{\ominus}{\text{Br}}$$

It will be seen that the oxygen atom, which is the negative end of the carbonyl dipole, behaves as an intramolecular nucleophile and attacks carbon as the CH_3S— group is leaving. Since methionine is one of the rarer amino acids in proteins, only a small number of fragments is obtained. For example, trypsinogen contains two methionyl residues in a linear sequence of 229 amino acids and is cleaved into the expected three fragments by cyanogen bromide.

Selective cleavage of the peptide bonds involving the carbonyl groups of tyrosine and tryptophan has been accomplished by oxidation with *N*-bromosuccinimide. For tyrosyl peptides, 3 moles of reagent are required for the cleavage of each peptide bond. The reaction (XLI → XLIV) involves an initial bromination, a concerted nucleophilic attack by carbonyl-oxygen and displacement of bromide ion, and hydrolysis of the imine (XLIII) to give a spirodienone lactone (XLIV):

O=C—NHR'
HO—
CH₂—CH.NH.CO.R
(XLI)
→
Br
Br—O—
O=C—NHR'
CH₂—CH.NH.CO.R
Br
(XLII)
↓
Br
O—C=NHR' Br$^{\ominus}$
O=
CH₂—CH.NH.CO.R
Br
(XLIII)
+ $R'\overset{\oplus}{N}H_3\overset{\ominus}{Br}$ ←H_2O—
Br
O—C=O
O=
CH₂—CH.NH.CO.R
Br
(XLIV)

The reaction is thus mechanistically related to the above procedure for cleaving methionyl bonds. For tryptophyl peptides, about 1·5 moles of *N*-bromosuccinimide are required for the cleavage of each peptide bond. The mechanism of cleavage (XLV → XLVIII) is similar in principle and gives ultimately a spirodioxindole-γ-lactone (XLVIII):

O=C.NH.R'
CH.NH.CO.R
CH₂
N
H
(XLV)
→
O=C—NH.R'
CH.NH.CO.R
CH₂
Br ⊕
N
H
H
(XLVI)
↓
O—C=$\overset{\oplus}{N}$H.R'
CH.NH.CO.R
CH₂
Br
N
H
H
(XLVII)
←
1. − HBr
2. + [O]
3. H_2O
O—C=O + $R'.\overset{\oplus}{N}H_3\overset{\ominus}{Br}$
CH.NH.CO.R
CH₂
N
H
O
(XLVIII)

Although the same reagent is employed, tryptophyl bonds are cleaved in preference to tyrosyl bonds and the method can be made quite selective.

Specific chemical methods for the cleavage of peptide bonds involving other amino acids have been developed using model compounds (Witkop, 1961), but have not been widely applied to the study of proteins.

2.8. Selective enzymic methods of cleaving peptide bonds. The use of aminopeptidases and carboxypeptidases for the selective removal of terminal amino-acid residues from peptides and proteins has already been discussed. This section deals with the use of other proteolytic enzymes which selectively rupture peptide bonds along the chain. Very frequently, native proteins are rather resistant to attack by proteolytic enzymes. Consequently, the proteins are usually denatured by heating or are subjected to a suitable procedure for cleaving disulphide bonds (§**2.4**) before exposure to proteolytic enzymes. The pancreatic enzyme, trypsin, selectively cleaves peptide bonds involving the carbonyl groups of the basic amino acids, lysine and arginine; these amino acids therefore become *C*-terminal in the fragments formed by enzymic cleavage (XLIX → L; LI → LII):

$$\underset{\text{(XLIX)}}{\text{R.CO.NH.}\overset{\overset{\oplus}{\text{NH}_3}}{\overset{|}{\underset{|}{(\text{CH}_2)_4}}}\text{CH.CO.NHR}'} \xrightarrow{\text{H}_2\text{O/trypsin}} \underset{\text{(L)}}{\text{R.CO.NH.}\overset{\overset{\oplus}{\text{NH}_3}}{\overset{|}{\underset{|}{(\text{CH}_2)_4}}}\text{CH.CO.O}^{\ominus}} + \text{R}'\overset{\oplus}{\text{N}}\text{H}_3$$

$$\underset{\text{(LI)}}{\text{R.CO.NH.}\overset{\text{NH}_2\text{—C}=\overset{\oplus}{\text{N}}\text{H}_2}{\overset{|}{\underset{|}{\text{NH}}}}\!\!\overset{}{\underset{|}{(\text{CH}_2)_3}}\text{CH.CO.NHR}'} \xrightarrow{\text{H}_2\text{O/trypsin}} \underset{\text{(LII)}}{\text{R.CO.NH.}\overset{\text{NH}_2\text{—C}=\overset{\oplus}{\text{N}}\text{H}_2}{\overset{|}{\underset{|}{\text{NH}}}}\!\!\underset{|}{(\text{CH}_2)_3}\text{CH.CO.O}^{\ominus}} + \text{R}'\overset{\oplus}{\text{N}}\text{H}_3$$

The action of trypsin on a protein normally yields one more peptide than the total number of lysine and arginine residues present, but in the event that one of these amino acids is *C*-terminal in the protein, the number of possible products is equal to the total number of lysine and arginine residues.

Trypsin is most active in the pH range 7–9 but is fairly rapidly inactivated by self-digestion. Addition of a low concentration of Ca^{2+} ions markedly inhibits self-digestion. Most preparations of trypsin contain traces of chymotrypsin which has a different specificity, but it is possible to inhibit the latter enzyme selectively and irreversibly by treatment with L-1-chloro-4-phenyl-3-toluene-*p*-sulphonamido-2-butanone (LIII) (see p. 136).

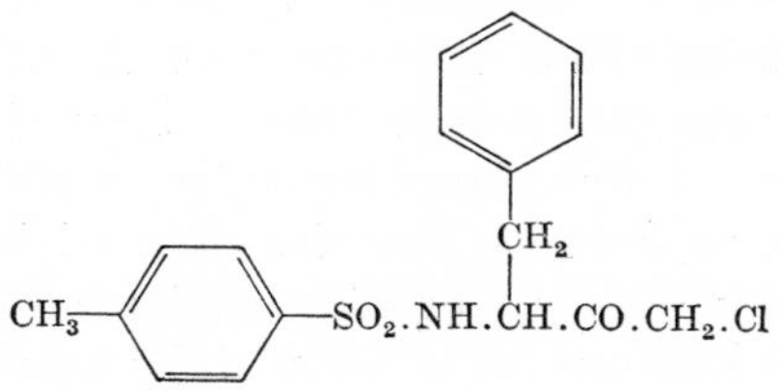

$$CH_3-C_6H_4-SO_2.NH.CH(CH_2C_6H_5).CO.CH_2.Cl$$

(LIII)

It is possible both to extend and restrict the action of trypsin on proteins. If a protein contains cysteine residues, these can be converted into *S*-(β-aminoethyl) cysteine groups (LIV) by reaction with ethylene-imine:

$$\text{R.CO.NH.CH(CH}_2\text{SH).CO.NHR}' \xrightarrow{(CH_2)_2\overset{\oplus}{N}H_2} \text{R.CO.NH.CH(CH}_2\text{S(CH}_2)_2\overset{\oplus}{N}H_3\text{).CO.NHR}' \xrightarrow{H_2O/\text{trypsin}} \text{R.CO.NH.CH(CH}_2\text{S(CH}_2)_2\overset{\oplus}{N}H_3\text{).CO.O} + \text{R}'\overset{\oplus}{N}H_3$$

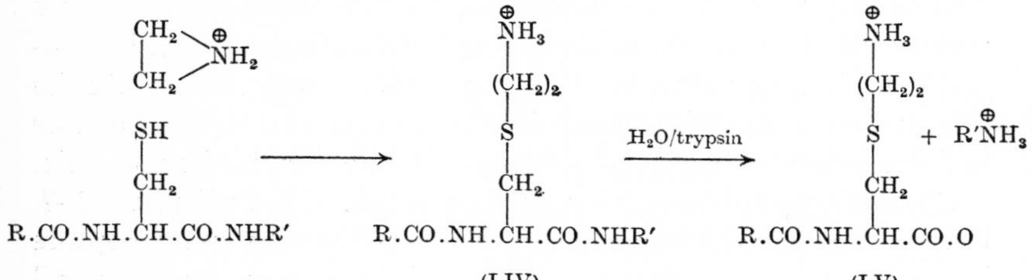

(LIV) (LV)

Since *S*-(β-aminoethyl) cysteine is an isostere of lysine (i.e. molecules of the two amino acids have approximately the same size and shape), it is not surprising that peptides of the former amino acid are hydrolysed by trypsin (LIV → LV). The same procedure is possible with cystine residues if the disulphide bonds are first reduced with mercaptoethanol (§**2.4**) before reaction with ethylene-imine.

If the ϵ-amino group of lysine is blocked, the action of trypsin can be restricted to the hydrolysis of arginyl peptide bonds (LI → LII). Several methods have been tried and perhaps the best involves trifluoroacetylation with ethyl thioltrifluoroacetate. After hydrolysis of the arginyl peptide bonds with trypsin and separation of the resulting peptides, trifluoroacetyl groups can be removed from the latter by treatment with aqueous piperidine at 0 °C. Each peptide that contains lysine can then be further degraded with trypsin.

Because of its high specificity, trypsin has been extensively used for comparing the primary structures of closely related proteins. The method, known as 'fingerprinting', involves a tryptic hydrolysis of the proteins and a comparison of the resulting peptides. Separation of these peptides is achieved by electrophoresis on filter paper followed by chromatography at right angles to the direction of electrophoresis. Peptides, located by spraying with ninhydrin, occupy a characteristic position on the 'fingerprint' map of spots. In this way, it was found for example, even before the primary structure of haemoglobin had been determined, that all except one of the tryptic peptides from sickle-cell haemoglobin were identical with those from normal human haemoglobin. (Sickle-cell anaemia results from an inborn error of haemoglobin synthesis and the name derives from the characteristic shape of the erythrocytes from persons suffering from this disease.) Elution and amino-acid analysis of the distinguishing peptide from each protein revealed that sickle-cell haemoglobin differed from normal haemoglobin by having valine in place of a glutamic acid residue. As the primary structure of haemoglobin became known it was possible to determine the locus of this structural variation in the molecule.

Chymotrypsin, like trypsin, is most active in the pH range 7–9, but is considerably less specific. It preferentially splits peptide bonds involving the carbonyl groups of tryptophan, tyrosine and phenylalanine, but it also hydrolyses more slowly peptide bonds involving the carbonyl groups of methionine, leucine, isoleucine and valine. Although preparations of chymotrypsin may contain traces of trypsin, the latter can be selectively inhibited with one of a series of naturally-occurring proteins such as soya-bean trypsin inhibitor. Hydrolysis of a protein with trypsin and with chymotrypsin in separate experiments, followed by a determination of the primary structures of the fragments, usually gives overlapping sequences which reveal the order in which the peptide sequences occur in the original protein (cf. p. 21). Chymotrypsin is widely used also for further degrading tryptic peptides to smaller fragments.

Other proteolytic enzymes, which are useful for degrading proteins, include pepsin (active at pH 1–2), papain (active at pH 5–7), and a bacterial enzyme, subtilisin (active at pH 6–8) from *Bacillus subtilis*. All these enzymes have broad specificities and are mainly used for the further degradation of large fragments which have been produced by tryptic hydrolysis or by one of the specific chemical methods of cleaving peptide bonds.

2.9. Mass spectrometry. Peptides are not sufficiently volatile to be studied directly by mass spectrometry, but this technique is feasible if the amino group(s) and carboxyl group(s) are blocked by acylation and esterification. The principal points of fragmentation are the bonds on either side of carbonyl groups:

$$\mathrm{R.CO.NH.CHR_1.CO \overset{\downarrow}{.} NH.CHR_2.CO}\overset{\downarrow}{} \ldots\ldots\ldots \mathrm{NH.CHR_{\mathit{n}}.CO \overset{\downarrow}{.} OCH_3}$$

$$\mathrm{R.CO.NH.CH\overset{\downarrow}{R_1}.CO.NH.CH\overset{\downarrow}{R_2}.CO} \ldots\ldots\ldots \mathrm{NH.CH\overset{\downarrow}{R_{\mathit{n}}}.CO.OCH_3}$$

In addition, there is a tendency for side-chains of valine, leucine, asparagine, serine, threonine and cysteine to be lost, but the masses of these fragments are known and can be allowed for in the mass spectrum. Analysis of the mass spectrum thus frequently permits the elucidation of the primary structure of a peptide. For example, *N*-decanoylalanylvalylglycylleucine methyl ester (mass = 526) gives intense peaks of mass 382, 325, 226 and 155 arising from the decanoylalanylvalylglycyl, decanoylalanylvalyl, decanoylalanyl and decanoyl ions respectively. Several other less intense peaks in the mass spectrum can be accounted for on the basis of fragmentation of amino acid side-chains as indicated above.

Mass spectrometry is limited to the study of small peptides at present. The low volatility of derivatives of large peptides and the operational range of mass spectrometers are obvious limitations. Nevertheless, the structure of fortuitine, an eicosanoyl-nonapeptide methyl ester (molecular weight = 1359) from *Mycobacterium fortuitum*, has been completely worked out by mass spectrometry.

2.10. Location of disulphide linkages. Strictly speaking, the presence of disulphide linkages affects the secondary and tertiary structures of proteins, but the experimental determination of their location in the molecule employs methods related to those used for the elucidation of the primary structure and is conveniently dealt with here.

Disulphides undergo exchange reactions rather easily and this behaviour makes the unambiguous location of disulphide linkages difficult in some cases. For example, in alkaline solution (e.g. during hydrolysis of proteins with trypsin or chymotrypsin) mercaptide ions and disulphides can undergo the following exchange reactions:

$$R.S^{\ominus} + R'.S.S.R' \rightleftharpoons R.S.S.R' + R'.S^{\ominus}$$
$$R.S^{\ominus} + R.S.S.R' \rightleftharpoons R.S.S.R + R'.S^{\ominus}$$

Such a situation can arise when a protein contains both cysteine and cystine. Even a catalytic amount of mercaptide ion, however, can catalyse exchange between two disulphides:

$$R.S.S.R + R'.S.S.R' \rightleftharpoons 2R.S.S.R'$$

These exchange reactions are inhibited by any of the reagents for thiol groups (pp. 25–26).

Exchange reactions also occur in strongly acid solution (e.g. during partial acid hydrolysis of proteins), but by a different mechanism:

$$R.S.S.R + H^{\oplus} \rightleftharpoons R.S^{\oplus} + R.SH$$
$$R.S^{\oplus} + R'.S.S.R' \rightleftharpoons R.S.S.R' + R'.S^{\oplus}$$
$$R'.S^{\oplus} + R.S.S.R \rightleftharpoons R.S.S.R' + R.S^{\oplus}$$

In this case, addition of a thiol reverses the first of the equilibria and inhibits exchange.

Determination of the positions of disulphide linkages in a protein is usually performed after the sequence of amino acids is known. The protein, with intact disulphide linkages, is partially hydrolysed using enzymes such as pepsin, trypsin or chymotrypsin and with precautions to minimize disulphide exchange reactions. The peptides are separated and none should contain more than one disulphide linkage. The disulphide linkage in each peptide is then cleaved and the two fragments are separated. Usually, analysis of amino acids and a partial determination of sequences serves to identify the portion of the original chain from which each fragment was derived. The technique can be illustrated by Sanger's work on the structure of insulin. Insulin consists of two polypeptide chains which can be separated after oxidation of the hormone with performic acid. The oxidized *A* chain contained four residues of cysteic acid while the oxidized *B* chain contained two residues of cysteic acid. In native insulin, therefore, there must be one disulphide linkage forming a loop in the *A* chain and two inter-chain disulphide linkages (§**6.8**). Insulin was hydrolysed with chymotrypsin in the presence of *N*-ethylmaleimide to prevent exchange reactions. One of the peptides, which contained cystine, gave two peptides containing cysteic acid after oxidation. The amino acid composition of one of these identified it as the *C*-terminal heptapeptide sequence of the *A* chain; the other, since it contained the only arginine residue, originated from near the *C*-terminus of the *B* chain. The site of the other inter-chain disulphide linkage was located by a similar method. If a peptide contains an intra-chain disulphide linkage with an intact peptide chain between the two halves of a cystine residue, it will not give rise to two fragments when it is oxidized with performic acid.

An elegant variation of the technique using performic acid oxidation has been devised. The mixture of peptides, some containing cystine, resulting from the hydrolysis of the original protein with a suitable enzyme such as pepsin, is separated by paper electrophoresis. The sheet of paper is then exposed to performic acid vapour and electrophoresis under the same conditions as before is carried out at right angles to the direction of the original separation. Peptides which do not contain cystine are unaffected by performic acid and appear on a diagonal. Those peptides which contain cystine, on the other hand, are each oxidized to two peptides containing cysteic acid and these give two spots off the diagonal during the second electrophoretic separation. The peptides containing cysteic acid can be eluted and their amino acid composition and sequence can be determined.

3. The Macromolecular Properties and Structures of Proteins

3.1. Introduction. In the previous chapter, methods were described for determining the primary structure of proteins. We shall now consider methods for studying the secondary, tertiary and quaternary structures of proteins, and the relationships between these and the primary structure.

The secondary and tertiary structures of a protein are influenced by the molecular size. For example, small peptides are probably free to assume many conformations in solution, while larger polypeptides have a molecular shape which is constrained by numerous noncovalent bonds. The biological activity of those proteins which are enzymes depends on the maintenance of a specific conformation. Rupture of more than a a limited number of noncovalent bonds leads to protein denaturation (§**3.10**).

3.2. Molecular weights of proteins. Estimates of the *minimum* molecular weight of a protein can be obtained from a total quantitative analysis of the amino acids it contains. There must obviously be an integral number of residues of each amino acid and a minimum molecular formula can be calculated in the same way as an empirical formula is calculated from the elementary analysis of a compound. Similarly, quantitative determination of *N*-terminal residues of amino acids in a protein can often give useful information about molecular size, but there are many examples of the *N*-terminal group being blocked by, for example, an acetyl group. Analysis of *N*-terminal groups is not possible in these cases. In addition, a protein molecule may consist of several chains which are held together only by noncovalent forces. *N*-Terminal analysis would lead, in this case, to a molecular weight which would be only a fraction of the true value. It is essential, therefore, to use physical methods to determine the molecular weight of a protein.

The molecular weight of a single molecular species in solution is independent of the method used for its determination. When protein molecules contain several polypeptide chains, however, equilibria may be established between several species such as monomers, dimers, trimers and so on. Such systems are called polydisperse and the value of the molecular weight obtained will depend on the method used for its determination. A number-average molecular weight, M_n, will result from measurements of colligative properties, where

$$M_n = \Sigma n_i M_i / \Sigma n_i$$

and the solution contains n_i molecules per unit volume of the ith species of molecular weight, M_i. On the other hand, techniques such as light-scattering and sedimentation in the ultracentrifuge measure the weight-average molecular weight, M_ω, where

$$M_\omega = \Sigma c_i M_i / \Sigma c_i$$

and c_i is the weight concentration per unit volume of the ith species. Since $c_i = n_i M_i$,

$$M_\omega = \Sigma n_i M_i^2 / \Sigma n_i M_i$$

and

$$\frac{M_\omega}{M_n} = \frac{\Sigma n_i \Sigma n_i M_i^2}{(\Sigma n_i M_i)^2} \geqslant 1$$

M_ω/M_n is unity only when the solution is monodisperse, i.e. it contains only a single molecular species. The value of the ratio, M_ω/M_n is a measure of the polydispersity of a solution.

Osmometry is a common method of determining M_n(Adair, 1961). For an ideal solution,

$$\frac{\Pi}{c} = \frac{RT}{M_n},$$

where Π is the osmotic pressure and c is the concentration of the protein in solution. (If Π is in dynes cm^{-2}, c should be in moles cm^{-3} and R in ergs mole^{-1} °C^{-1}.) Consequently, Π/c should be independent of c. In practice, a plot of Π/c versus c is frequently found to be linear with a non-zero slope. This is attributed to non-ideal behaviour. By analogy with the van der Waals equation for imperfect gases, it is possible to derive the equation:

$$\frac{\Pi}{c} = \frac{RT}{M_n} + \frac{BRTc}{M_n^2},$$

where B is known as the second virial coefficient and is usually positive. If association of the protein occurs, however, Π/c decreases in a nonlinear fashion with increasing values of c. Typical plots are illustrated in fig. 3.1. In either case, determination of molecular weight involves the measurement of the osmotic pressure over a range of values of c and extrapolation to zero concentration.

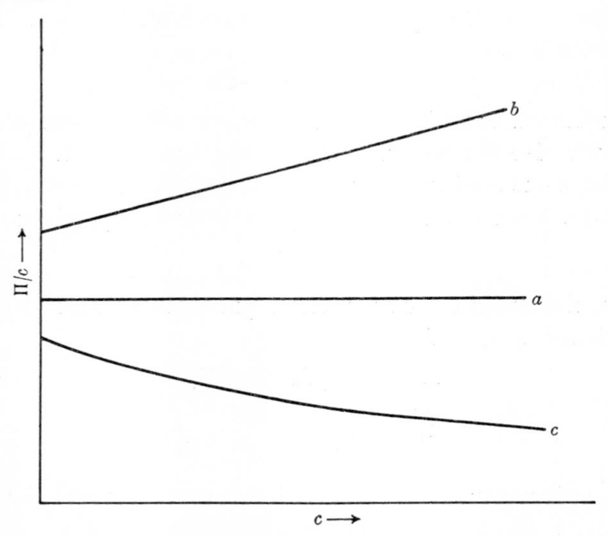

Fig. 3.1. Variation of Π/c with c. The linear curve a, with zero slope, represents ideal behaviour. Non-ideal behaviour is represented by curve b, which has a slope equal to BRT/M_n^2. When association occurs, a curve of negative slope similar to c is obtained. In each case, the ordinal intercept is equal to RT/M_n.

There is an additional complication in determining molecular weights by osmometry. If the protein has a net charge, the concentrations of diffusible ions will not be identical on each side of the membrane. There will be an excess of ions of the opposite sign to that of the protein on the same side of the membrane as the protein. This behaviour, known as the Donnan effect, leads to increased osmotic pressures. The Donnan effect may be diminished by working at or near the isoelectric point or at high ionic strengths.

The development of the ultracentrifuge by Svedberg in 1923 has provided perhaps the most popular method for the determination of the molecular weight of a protein. At the speeds attainable in the ultracentrifuge (e.g. 60 000 r.p.m.), the centrifugal force applied to large molecules is sufficient to overcome diffusion. The macromolecules sediment and present a boundary which can be observed and analysed by suitable optical methods (Claesson & Moring-Claesson, 1961). There are several methods for determining molecular weights based on sedimentation behaviour.

In the sedimentation-velocity method, the velocity of migration of the boundary, $\mathrm{d}r/\mathrm{d}t$, is determined at a distance, r, from the axis of rotation at an angular velocity, ω. The sedimentation coefficient, s, is defined by the equation:

$$s = \frac{\mathrm{d}r}{\mathrm{d}t}\frac{1}{r\omega^2}.$$

The molecular weight is then given (for an ideal solution) by the equation:

$$M_\omega = \frac{RTs}{D(1-\bar{v}\rho)},$$

where D is the diffusion coefficient, $\bar{v}$ is the partial specific volume and ρ is the density of the solution. The diffusion coefficient must be separately determined (Svensson & Thompson, 1961) and the partial specific volume may be obtained either experimentally or by computation from the amino acid content of the protein (Cohn & Edsall, 1943). As a method for determining molecular weights, the sedimentation-velocity approach is less convenient than others described below because of the additional experimentation involved to obtain D and possibly $\bar{v}$. Determination of the diffusion coefficient may be justified, however, in another connection. The frictional coefficient of a particle in solution, f, is given by the equation:

$$f = \frac{RT}{ND},$$

where N is Avogadro's number. For an anhydrous spherical molecule,

$$f_0 = 6\pi\eta r_0,$$

where η is the coefficient of viscosity of the medium and r_0 is the radius of the molecule. If the molecular weight can be determined by a method which does not involve a knowledge of the diffusion coefficient, f_0 may be calculated from the equation:

$$f_0 = 6\pi\eta\left(\frac{3\bar{v}M_w}{4\pi N}\right)^{\frac{1}{3}}.$$

The frictional ratio, f/f_0, is greater than unity owing to solvation of the protein molecule or deviation from spherical shape. The observed frictional ratio may be considered to derive from the product of two ratios, $(f/f_0)_s$ due to solvation and $(f/f_0)_a$ due to asymmetry of the molecule

$$f/f_0 = (f/f_0)_s(f/f_0)_a$$

$(f/f_0)_s$ may be obtained approximately by determining the weight of solvent, w, of density, ρ, which is bound per gram of protein. Then,

$$(f/f_0)_s = \left(1 + \frac{w}{\bar{v}\rho}\right)^{\frac{1}{3}}.$$

Values of about 1·1 are common.

Values of f/f_0 for simple asymmetric particles such as ellipsoids of revolution have been calculated theoretically (fig. 3.2). It will be seen that an estimate of the axial ratio is possible if the frictional ratio and degree of solvation are known.

The ultracentrifuge can be used in other ways to determine the molecular weight of a macromolecular solute. If the angular velocity is rather less than that used in the sedimentation-velocity method, it is possible to obtain conditions such that equilibrium is established between the tendency to sediment and the opposing tendency for diffusion to occur. Under such conditions, the following equation obtains:

$$M_w = \frac{RT}{(1-\bar{v}\rho)\,\omega^2 rc}\frac{\mathrm{d}c}{\mathrm{d}r} = \frac{2RT\ln(c_2/c_1)}{\omega^2(1-\bar{v}\rho)\,(r_2^2 - r_1^2)},$$

where c_1 and c_2 are the concentrations of the solute at distances r_1 and r_2 from the axis of rotation. It will be noticed that the partial specific volume is still required but not the diffusion coefficient. Equilibrium may take a considerable time to be established and a method has been devised by Archibald in which the molecular weight can be determined from measurements obtained before equilibrium is established. Since there is no net flow of solute through the meniscus of the solution and through the bottom of the cell, the conditions which exist at equilibrium throughout the cell, apply at all times to the meniscus and the bottom of the cell. If c_m, the concentration of solute at the meniscus, is determined and if dc/dr is measured over a region and extrapolated to the

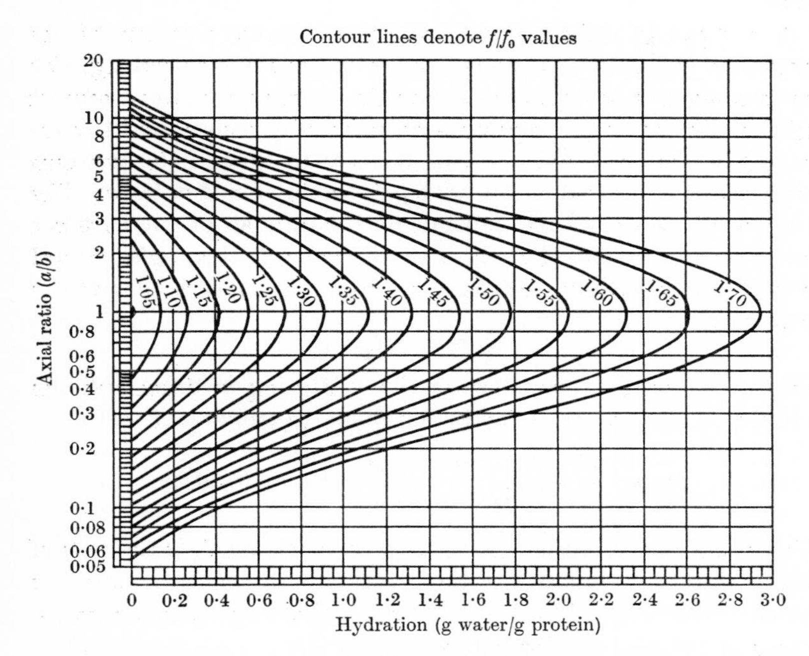

Fig. 3.2. Contour plots of the frictional ratio f/f_0. [J. L. Oncley (1941). *Ann. N.Y. Acad. Sci.* **41**, 121.]

value $(dc/dr)_m$ at the meniscus, the molecular weight can be calculated from the above equation used for the sedimentation-equilibrium method. Similarly, a value of the molecular weight can be obtained from measurements of c and dc/dr at the bottom of the cell. The two values of M_w may be different and time-dependent due to polydispersity or the existence of equilibria between monomer and polymers.

In a new application of the ultracentrifuge, a protein solute can be sedimented in a sucrose solution whose concentration exhibits a gradient, rising from the top to the bottom of the cell. A band is formed where the density of the sucrose solution corresponds to the density of the macromolecule and the associated components in solution. The band width is dependent on the diffusion and polydispersity of the molecules. For monodisperse solutes, the molecular weight can be determined from the shape and width of the band.

A completely different method for the determination of the molecular weight of a protein depends on the scattering of light by macromolecules in solution. A parallel, monochromatic beam of light is allowed to impinge on a solution of a protein and the intensity of the scattered light, i_θ, relative to that of the incident beam, I_0, is measured at an angle, θ, to the incident beam. The amount of light scattered depends on the number of particles per unit volume, the wavelength of the light used, the angle, θ, and several other factors. Rayleigh's ratio, R_θ, is defined by the equation:

$$R_\theta = i_\theta r^2/I_0,$$

where r is the distance between the scattering medium and the detector of the scattered light. For small, isotropic molecules,

$$R_\theta = \frac{2\pi^2 n_0^2(1+\cos^2\theta)}{N\lambda_0^4}\left(\frac{\mathrm{d}n}{\mathrm{d}c}\right)^2 Mc = K_\theta Mc.$$

Where n_0 is the refractive index of the solvent, $\mathrm{d}n/\mathrm{d}c$ is the rate of change of refractive index of the solution with concentration, c is the concentration, N is Avogadro's number and λ_0 is the wavelength of the incident light in a vacuum.

When the solute molecules are larger than about $0{\cdot}05\lambda_0$, the above equation must be modified by including a particle-scattering factor, P_θ, to allow for interference between beams of light scattered at different points on the particle:

$$R_\theta = K_\theta P_\theta Mc.$$

The value of P_θ depends on the shape of the molecule, but is unity for all shapes when θ is zero. If both θ and c tend to zero:

$$\frac{K_\theta c}{R_\theta} = \frac{1}{M}\left[1 + \frac{16\pi^2 R_G^2 n_0^2 \sin^2\theta/2}{3\lambda_0^2}\right],$$

where R_G is the root mean square average radius of gyration of the molecule. Both M and R_G can be determined by measuring R_θ at different angles and over a range of concentrations. In a graphical double extrapolation, known as the Zimm plot, $K_\theta c/R_\theta$ is plotted against $\sin^2\theta/2+kc$, where k is an arbitrary constant chosen to provide a convenient spread of the data. When the double extrapolation to zero angle and concentration has been carried out, the molecular weight and R_G can be determined from the equations:

$$M = 1/\text{ordinal intercept},$$

$$R_G^2 = \frac{3\lambda_0^2}{16\pi^2 n_0^2}\cdot\frac{(\text{slope})_{c=0}}{\text{ordinal intercept}}.$$

The value of R_G^2 is related to the dimensions of molecules of simple shape. For spheres,

$$R_G^2 = 3r^2/5,$$

where r is the radius of the sphere. For rods,

$$R_G^2 = l^2/12,$$

where l is the length of the rod. For random coils,

$$R_G^2 = \overline{h}^2/6,$$

where $(\overline{h}^2)^{\frac{1}{2}}$ is the root mean square average distance between the ends of the coil. If independent information concerning the shape or dimensions of the molecule is available, light-scattering measurements provide a test of the proposed model.

3.3. The geometry of the peptide bond. X-ray crystallographic studies have shown that the peptide bond preferentially adopts the planar, *trans* configuration (fig. 3.3). The bond between the nitrogen and carbonyl-carbon atoms is considerably less than 1·47 Å which is the usual length of a C—N bond. It must be assumed that the bond has partial double bond character due to resonance. This accounts for the planarity of the peptide bond, but there is an alternative approach. In the carbonyl group, the carbon atom is hydridized in the sp^2 state with three planar σ-bonding orbitals at 120°. There is a p-orbital normal to the plane of the sp^2 hybrids. The carbonyl bond consists of (i) a σ-bond formed by overlapping one of the sp^2 hybrids with a p-orbital of oxygen, (ii) a π-bond formed by overlapping the p-orbital of carbon with the p-orbital of oxygen. In the planar amide group, overlap between the carbon π- and nitrogen p-orbitals occurs and this is another way of describing the resonance between the two canonical forms depicted in fig. 3.3. Rotation about the C—N bond would preclude this overlap and result in a loss of resonance energy amounting to perhaps 10 kcal. mole^{-1}. *Cis*-peptide bonds are found in small cyclic peptides such as 1,4-dioxopiperazines, where a *trans* peptide bond could not exist. Generally, however, the *trans* configuration seems to be energetically favoured.

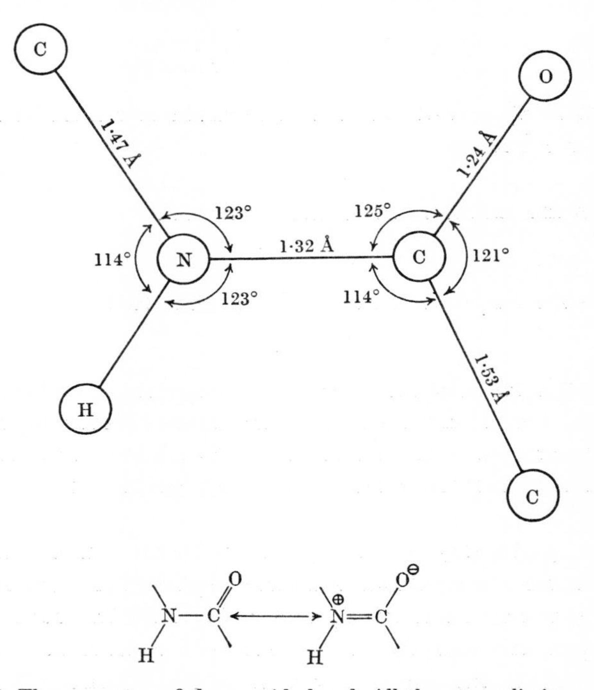

Fig. 3.3. The geometry of the peptide bond. All the atoms lie in one plane.

3.4. Noncovalent bonds in proteins. A hydrogen bond can exist when two electronegative atoms, one of which is bonded to hydrogen, can approach one another closely with the hydrogen atom preferably linearly placed between them. The hydrogen end of a bond between hydrogen and an electronegative atom carries a fractional positive charge and is attracted towards an unshared pair of electrons in another electronegative atom:

$$\mathrm{A}\text{——}\mathrm{H}\cdots\cdots \odot\, \mathrm{B}$$

If the two electronegative atoms involved in the formation of a hydrogen bond are identical, the hydrogen atom is shared equally between them. In any case, when hydrogen bonding occurs, the distance between the two electronegative atoms concerned is less than the sum of their van der Waals radii. Although hydrogen bonds have low bond energies of the order of 5 kcal. $mole^{-1}$, there are so many groups capable of forming them in proteins (table 3.1) that they play a large part in determining the molecular conformation.

TABLE 3.1

Hydrogen-donors	Acceptors
Amide nitrogen	Amide carbonyl oxygen
Carboxyl	Carboxyl
Hydroxyl (aliphatic and aromatic)	Carboxylate anion
Ammonium ion	Hydroxyl (aliphatic and aromatic)
Guanidinium ion	Phenolate anion
Imidazolium ion	Amino
Thiol	Guanidine
	Imidazole
	Thiolate anion

The ϵ-amino groups of lysyl residues and guanidinium groups of arginyl residues on the one hand and the β- and γ-carbonyl groups of aspartyl and glutamyl residues on the other can form salt linkages of the type:

$$- \overset{\oplus}{N}H_3 \quad \overset{\ominus}{} \left\{ \begin{matrix} O \\ O \end{matrix} \right\rangle C -$$

Such bonds have a free energy of formation of approximately 10 kcal. $mole^{-1}$. Salt linkages would not be expected to survive in the presence of high concentrations of electrolytes. Since salts do not readily denature proteins, it may be concluded that salt linkages do not play an essential role in determining protein conformation.

Polypeptide chains tend to fold in such a way that the nonpolar side chains of amino acids such as leucine, isoleucine, valine, phenylalanine and tryptophan tend to come together. The term 'hydrophobic bond' has been coined to describe this type of noncovalent interaction although it tells us nothing of its nature. Van der Waal's forces between atoms are likely to be relatively unimportant except when very close approach is possible, since the energy of bonding for this type of interaction is inversely proportional to the sixth power of the distance between the atoms involved. Secondly, overlapping of π-orbitals may occur when a suitable steric relationship exists between amino acids with aromatic side chains. This type of bonding is also likely to be of little consequence in determining the conformation of protein molecules, since many hydrophobic bonds involve amino acids with aliphatic side chains. The most important source of hydrophobic bonding appears to stem from the interaction between apolar side chains of amino acids and water molecules. Water molecules in the vicinity of apolar groups possess a greater degree of organization than in the bulk of the solvent. The water molecules adjacent to the apolar structure have limited possibilities for forming hydrogen bonds with other water molecules and they therefore assume a regular arrangement in order to achieve the maximum degree of hydrogen bond formation. If two or more apolar groups can come into sufficiently close contact, they exclude water molecules and form a kind of intramolecular micelle. The number of water molecules which have a highly organized structure near a cluster of apolar groups will be less than the number of water molecules which would be structurally organized around the several apolar groups if the latter were separated. In other words, the entropy of the solvent is increased if apolar groups come together and this is the thermodynamic driving force which favours the formation of hydrophobic bonds. It is not surprising that protein molecules tend to be folded in such a way that the maximum number of hydrophobic bonds are formed in the *interior* of the molecule and amino acids with hydrophilic side chains tend to occur on the surface of the molecule. Addition of organic solvents to aqueous solutions of proteins tends to destroy the structural organization of water and therefore nullifies the thermodynamic driving force tending to form hydrophobic bonds. The latter therefore tend to break down in the presence of organic solvents and the protein molecule is likely to lose its normal conformation and probably its biological activity.

3.5. Polypeptide conformations. Pauling and his colleagues examined plausible models of polypeptide structure based on considerations of potential energy. They argued that the most stable conformation would have (*a*) a planar peptide bond (§**3.3**); (*b*) the maximum number of hydrogen bonds of the type,

$$>C{=}O\cdots H{-}N<,$$

in which donor, acceptor and hydrogen atoms would not deviate by more than 30° from linearity; (*c*) bond angles and lengths similar to those in small molecules such as amides; (*d*) orientation about C—C and N—C single bonds close to the potential energy minima for rotation about these bonds. The most satisfactory structure, based on the above criteria, is the α-helix (fig. 3.4). Although the helix can have either a right-handed or left-handed sense, the former is energetically preferred.

The main structural features of the α-helix can be summarized as follows: (*a*) hydrogen bonds between carbonyl-oxygen and peptide-nitrogen atoms occur at intervals such that there are three complete amino acid residues between them

$$(\underset{\substack{\|\\ O}}{-C}-[NH.CHR.CO]_3-\underset{\substack{|\\ H}}{N}-) \quad O\cdots\cdots H$$

and the hydrogen bonds are almost parallel to the axis of the helix; (*b*) five turns of the helix contain eighteen amino acids (i.e. 3·6 residues per turn) and the pitch of the helix is 5·4 Å; (*c*) side chains of amino acids protrude from the helix; (*d*) the hole down the centre of the helix is too narrow to accommodate solvent molecules. The existence of the right-handed helix in proteins in the solid state has been proved by X-ray diffraction studies (§**3.9**) and there is less direct evidence of its presence in solution (§**3.8**).

The α-helix owes its stability to the formation of *intramolecular* hydrogen bonds. Extended conformations are also feasible if *intermolecular* hydrogen bonds are formed. There are two main types, the parallel and antiparallel pleated sheets (fig. 3.5). The former has all the *N*-termini oriented in the same direction, while the latter has alternate chains running in opposite directions. Those proteins which exist in extended conformations are usually fibrous and rather insoluble in water.

The foregoing structures represent limiting conformations, since there are several factors which disturb the α-helix and give rise to conformations consisting of alternate helical and randomly coiled

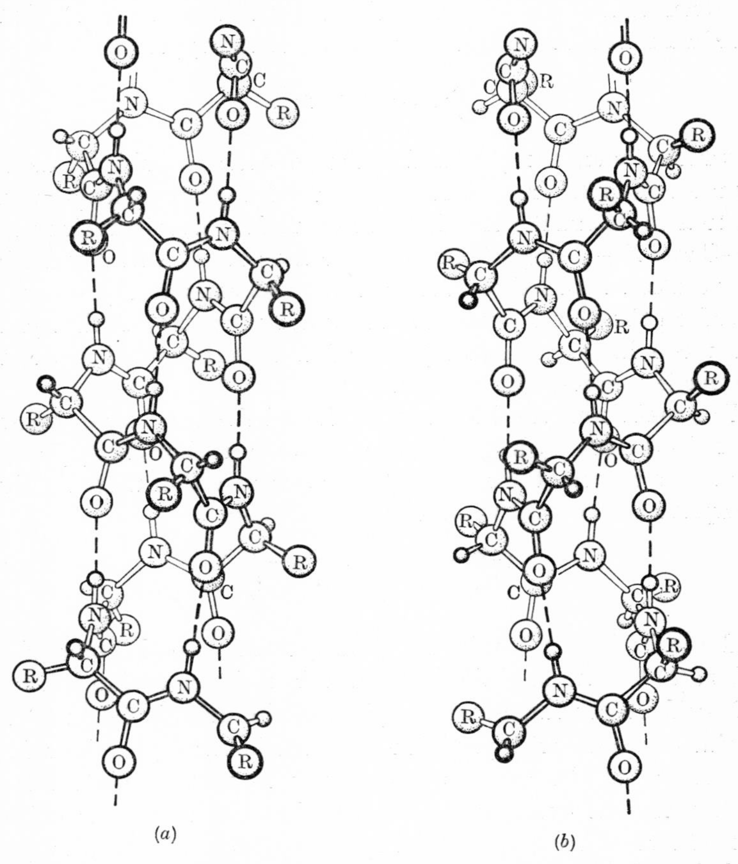

Fig. 3.4. The left-handed (*a*) and right-handed (*b*) α-helices constructed from L-amino acids. Notice the $\rangle N{-}H\cdots\cdots O{=}C\langle$ intramolecular hydrogen bonds. [From L. Pauling (1960). *The Nature of the Chemical Bond*, 3rd ed. Ithaca, N.Y.: Cornell University Press.]

segments. Although proline can exist in the α-helix, the bulky pyrrolidine ring interferes with helix formation by the amino acid residue which acylates the imino group of the pyrrolidine ring. Thus, proline residues usually mark breaks in the helical structure

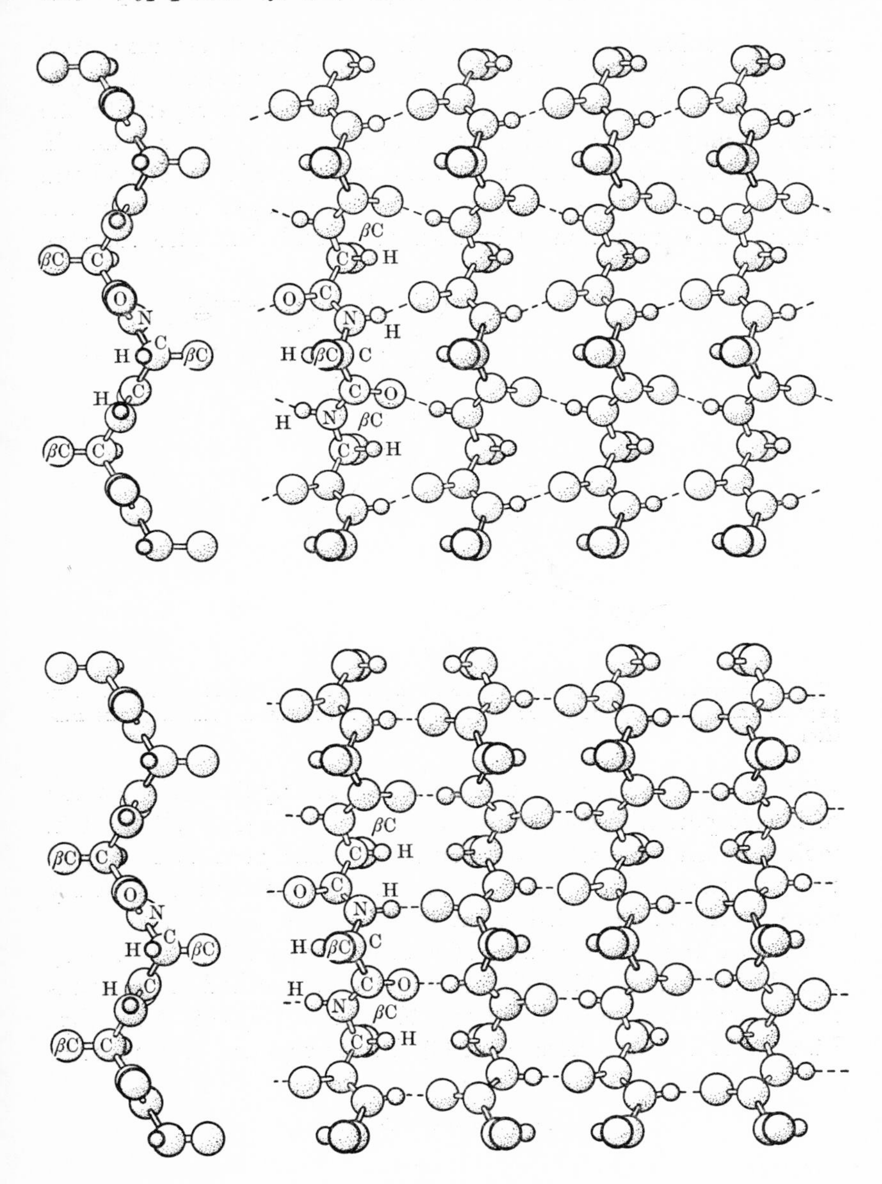

Fig. 3.5. Parallel (top) and antiparallel (bottom) pleated sheets. Notice the >N—H······O=C< intermolecular hydrogen bonds. [From L. Pauling (1960). *The Nature of the Chemical Bond*, 3rd ed. Ithaca, N.Y.: Cornell University Press.]

3-2

of proteins. The side chains of valine and isoleucine can also interfere sterically with helix formation. The formation of intramolecular and intermolecular disulphide bridges, especially the former, may prevent helix formation. In this case, the loss in potential energy resulting from the formation of a covalent bond compensates for the inability to form hydrogen bonds in the α-helix. The formation of hydrophobic bonds will also compete

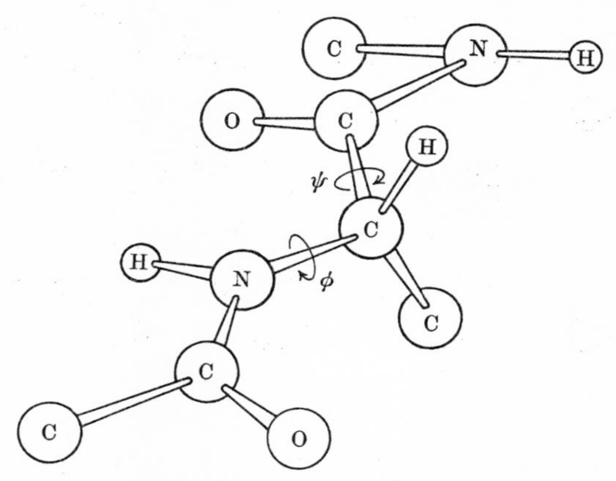

Fig. 3.6. Perspective drawing of a section of a polypeptide chain representing two peptide units. The peptide bonds are drawn in the most usual (planar) conformation.

with hydrogen bonding as a means of producing the conformation of lowest potential energy. As mentioned above, (§**3.4**), protein molecules tend to assume a conformation which permits the formation of hydrophobic bonds between apolar side chains, especially when these can be buried in the interior of the molecule.

Because the conformation of a protein molecule can be very complex, it is desirable to have some simple method of quantitative description (Edsall, Flory, Kendrew, Liquori, Némethy, Ramachandran & Scheraga, 1966). In the segment of a protein,

$$\text{—C}^{\alpha}\text{HR—}\overset{\overset{\displaystyle\text{O}}{\|}}{\text{C}'}\text{—}\underset{\underset{\displaystyle\text{H}}{|}}{\text{N}}\text{—C}^{\alpha}\text{HR—}$$

(referred to as a 'peptide unit'), there is usually no rotation about the C′—N bond because the peptide bond is planar. Consequently, it is possible to describe the conformation in terms of the rotations about the C^{α}—C′ bond (ψ) and about the N—C^{α} bond (ϕ) (fig. 3.6). The complete conformation of a protein containing n amino acid residues can thus be described by the parameters

$$(\phi_1, \psi_1)\ (\phi_2, \psi_2)\ \ldots\ (\phi_n, \psi_n).$$

Values of ϕ and ψ for some regular conformations are given in table 3.2.

TABLE 3.2. *Angles of rotation for some regular protein conformations*

	ϕ	ψ
Fully extended chain	0°	0°
Right-handed α-helix	132°	123°
Left-handed α-helix	228°	237°
Parallel pleated sheet	113°	293°
Antiparallel pleated sheet	145°	325°

3.6. Potentiometric titration of proteins. Potentiometric titration is a useful technique for studying the secondary and tertiary structures of proteins, since the existence of hydrogen bonds, hydrophobic bonds and salt-linkages can cause notable shifts of pK_a. Although proteins contain both acidic and basic groups, it is usual to consider the latter in the form of their conjugate acids and pK_a rather than pK_b values are used to describe their behaviour on titration.

A simple amino acid exists in aqueous solution as a dipolar ion (zwitterion), $\overset{\oplus}{N}H_3.CHR.CO.O^{\ominus}$. The carboxyl group has a pK_a value of approximately 2·3 and is therefore a stronger acid than say acetic acid ($pK_a \sim 4{\cdot}8$) due to the electrostatic effect of the $\overset{\oplus}{N}H_3$— group. A carboxylate ion might be expected to be electron-repelling. If this were the whole story, the carboxylate ion would inhibit the dissociation of a proton from the $\overset{\oplus}{N}H_3$—group and the pK_a of the latter would be expected to be higher for an amino acid than for a simple aliphatic amine ($pK_a \sim 10{\cdot}6$). In fact, the carbonyl group of the carboxylate ion exerts an opposing electron-attracting effect which apparently predominates since the pK_a of the $\overset{\oplus}{N}H_3$— group in α-amino acids is about 9·7. In proteins, only the electron-attracting effect of the peptide bond is operative and the pK_a of the $\overset{\oplus}{N}H_3$—drops to 7·4–7·9. Likewise, the terminal carboxyl group in a protein is influenced by the electron-attracting peptide linkage and it has a pK_a of 3·1–3·8, intermediate between that of acetic acid and an α-amino acid. The ϵ-amino groups in the side chains of lysyl residues are well insulated from the inductive effects due to peptide bonds and so have pK_a values near 10·2, only slightly lower than those of simple aliphatic amines. Similarly, β- and γ-carboxyl groups of aspartyl and glutamyl residues are separated from peptide bonds by a saturated aliphatic chain and are only a little stronger than acetic acid.

There are several other titratable groups in proteins. The imidazole ring in the side chain of histidine can be protonated to give the imidazolium ion (pK_a = 6·0–7·4); the thiol group in the side chain of cysteine can lose a proton (pK_a = 8·5–9·5); the phenolic hydroxyl of tyrosine has pK_a = 9·6–10·0, while the guanidino group of arginine is very basic (pK_a > 12) and is not normally titratable without denaturing a protein.

It is obvious that, in the pH range 4–10, there will be a large number of titratable groups and the interpretation of titration curves is a complex subject. In addition to the number of groups involved, their pK_a values can be perturbed by several factors. Firstly, when charged groups are in close proximity or when salts are present, pK_a values will be influenced by electrostatic effects. Titration thus gives apparent pK_a values and the intrinsic pK_a values have to be computed by applying a suitable correction factor based on the Debye–Hückel theory:

$$\mathrm{p}K = \mathrm{p}K_{\mathrm{intrinsic}} - 0{\cdot}868wZ,$$

where Z is the net charge (the algebraic difference between the numbers of positive and negative charges) and w is given by the equation:

$$w = \frac{Ne^2}{2DRT}\left(\frac{1}{b} - \frac{\kappa}{1+\kappa a}\right) = 3{\cdot}57\left(\frac{1}{b} - \frac{\kappa}{1+\kappa a}\right) \quad \text{in water at } 25^\circ.$$

In the last equation, N is Avogadro's number, e is the electronic charge, D is the dielectric constant of the medium, R is the gas constant, T is the absolute temperature, b is the radius of the charged protein molecule (Å) assuming that it is spherical, a is the distance of closest approach of a small ion and the protein sphere ($\sim b+2{\cdot}5$ Å), and κ is the Debye–Hückel parameter and is defined by the equation:

$$\kappa = \left(\frac{4\pi Ne^2}{DRT}\Sigma c_i Z_i^2\right)^{\frac{1}{2}},$$

where c_i and Z_i are the concentration and charge of the ith ionic species. At best, this correction is only approximate since it assumes that the protein molecule is spherical with the charges uniformly distributed over the surface. Again, the effective dielectric constant in the vicinity of the protein molecule is likely to be much lower than D, the bulk dielectric constant of the solvent.

In addition to electrostatic effects, the presence of hydrophobic bonds or hydrogen bonds will influence the apparent pK_a values of titratable groups. If a titratable group is 'buried' in a part of the protein molecule in which there is appreciable hydrophobic bonding, it will behave as if it were dissolved in an apolar solvent of low dielectric constant. Dissociation of a carboxyl group into carboxylate anion and proton occurs much less readily in a solvent of low dielectric constant than in water, since two charged ions are produced from an uncharged group. Consequently, the pK_a is increased. In contrast, the dissociation of a proton from a $\overset{\oplus}{N}H_3$— group results in no net change in the number of charged particles and consequently the pK_a value of such a group is relatively insensitive to a change of dielectric constant.

When hydrogen-bonding involves titratable groups, the pK_a may be increased or decreased according to circumstances. If the acidic form of an acid is acting as a donor, removal of the proton will be inhibited, and the pK_a will be increased. Conversely, if the basic form of a conjugate acid-base system is acting as an acceptor, addition of a proton to the base will be inhibited, and the pK_a will be lowered.

Frequently, the titration of a native protein results in abnormal shaped curves or the uptake of alkali may be time-dependent. For example, if the phenolic hydroxyl group of a tyrosyl residue functions as a donor in the formation of a hydrogen bond, its pK_a will be increased above the normal value of about 10·0. The expected consumption of alkali from about pH 9 would not be observed. Addition of alkali to about pH 10, however, might be sufficient to rupture the hydrogen bond. The liberated hydroxyl group would then be titrated and there would be a very steep rise in alkali consumption as the pH was further increased. Back-titration with acid from this high pH value would give a normal titration curve (fig. 3.7). Conversely, if a carboxylate anion, for example, functions

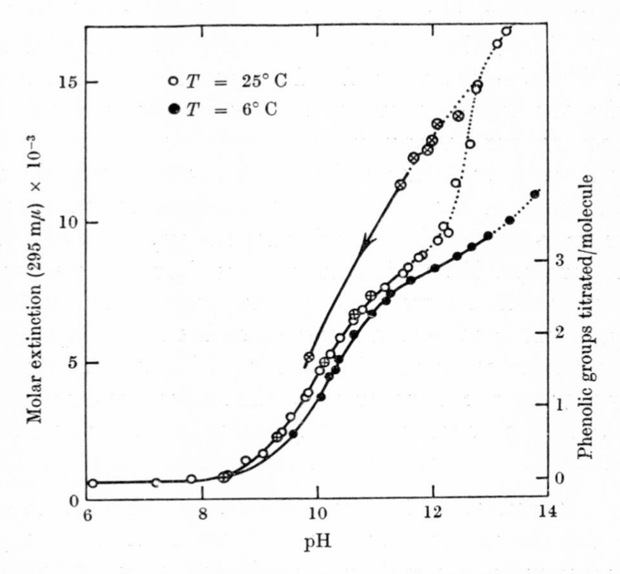

Fig. 3.7. Ionization of the hydroxyl groups of tyrosyl residues in bovine pancreatic ribonuclease. The ionization is independent of time where solid lines are drawn, but increases with time where dotted lines are drawn: ⊕, solutions at 25° reversed after exposure to pH 11·5; ⊗, solutions at 25° reversed after exposure to pH 12·7. Notice that the hydrogen bonds are less stable at the higher temperature. [C. Tanford, J. D. Hauenstein & D. G. Rands (1955). *J. Amer. chem. Soc.* **77**, 6409.]

as the acceptor in the formation of a hydrogen bond, its pK_a will be decreased. On titration with acid there would be a delay in the uptake of protons. If the hydrogen bond is ruptured at a pH value near or below the normal pK_a value of such a carboxylic acid, the liberated carboxylate anion would be titrated and the rate of uptake of protons with change of pH would be greater than expected. On back-titration with alkali, the carboxylic acid would titrate normally.

Although the interpretation of titration curves is a matter of some complexity, useful information can be obtained about the secondary and tertiary structures of proteins, especially when groups give rise to abnormal titration curves. Assistance in the interpretation of potentiometric titration curves can be obtained by the use of other techniques. For example, the total number of titratable groups of each kind can be ascertained from a complete analysis for amino acids. Secondly, the phenolate anion from tyrosine has an ultraviolet absorption spectrum which is different from that of tyrosine itself and so the dissociation of tyrosine hydroxyl groups can be separately followed by spectrophotometric titration. Finally, the number of protonated amino groups can be determined by titration in the presence of formaldehyde. Protons are removed from $\overset{\oplus}{N}H_3$— groups according to the following mechanism:

$$\overset{\oplus}{N}H_3\text{—} \rightleftharpoons H^{\oplus} + NH_2\text{—} \overset{H.CHO}{\rightleftharpoons} HO.CH_2.NH\text{—} \overset{H.CHO}{\rightleftharpoons} (HO.CH_2)_2N\text{—}$$

In the presence of a large excess of formaldehyde, the equilibria are almost completely displaced to the right. The liberated protons can thus be titrated with alkali.

More detailed accounts of the technique of potentiometric titration of proteins can be found in reviews by Tanford (1962) and Kenchington (1960).

3.7. Ultraviolet spectra of proteins. The presence of the peptide bond causes proteins to absorb light strongly in the wavelength range 180–220 mμ. Experiments with polypeptides containing only one species of amino acid have shown that the spectra of the α-helical and random coil forms are appreciably different. The extinction coefficient at 190 mμ is decreased when a a-helix is formed (hypochromic effect) and a new absorption band appears at about 205 mμ. It is possible to calculate the fraction of amino acid residues in helical conformation in a protein by determining the extinction coefficient at 190 mμ and correcting for absorption by the side-chains of amino acids. The results agree fairly well with those derived by measurements of optical rotatory dispersion (§**3.8**).

Some information concerning the tertiary structure as well as helical content of proteins can be obtained from ultraviolet spectrophotometric measurements. Proteins containing the aromatic amino acids, phenylalanine, tyrosine and tryptophan absorb in the 250–300 mμ region of the ultraviolet spectrum. Phenylalanine has the lowest values of $\lambda_{max.}$ (257 mμ) and $\epsilon_{max.}$ ($\sim$ 200) and its contribution to the spectrum of proteins is frequently swamped by the absorption due to tyrosyl and tryptophyl side-chains. The spectral contributions arising from tyrosyl and tryptophyl residues are

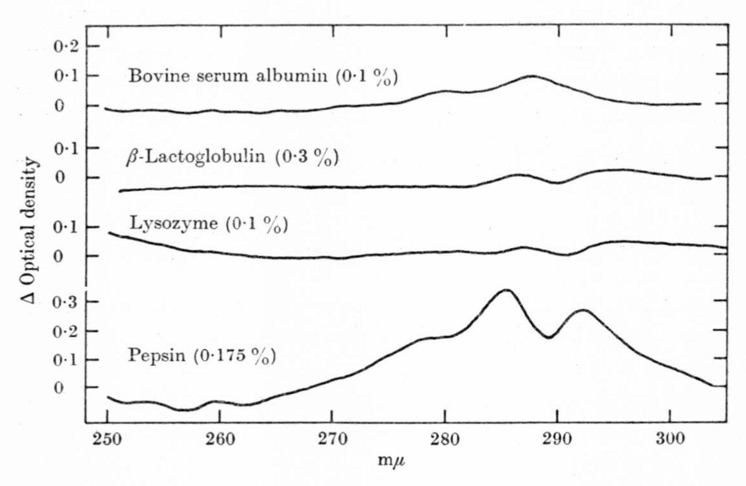

Fig. 3.8. Difference spectra of proteins resulting from acidification (prior denaturation in the case of pepsin). Native protein at pH 6–7 was placed in the sample beam; acidified protein at pH 1–1·5 (pH 2·2 for pepsin) was placed in the reference beam. [S. Yanari & F. A. Bovey (1960). *J. biol. Chem.* **235**, 2818.]

dependent on the environment of these amino acids in the protein molecule. A neighbouring charged group can influence the spectra of aromatic amino acids and the spectrum due to a tyrosyl residue is also altered if the phenolic hydroxyl group is involved in hydrogen-bonding. A more general cause of spectral perturbation, however, appears to be the formation of hydrophobic bonds. Model compounds such as indole and phenol absorb at a longer wavelength (red shift) in hydrophobic solvents of high refractive index than they do in hydrophilic solvents of low refractive index. When the side chains of aromatic amino acid residues in proteins participate in the formation of hydrophobic bonds, they are effectively dissolved in a nonpolar solvent of high refractive index. If the protein is denatured, hydrophobic bonds are ruptured and the wavelength of maximum absorption is shifted to lower wavelengths (blue shift). The spectral change is small and cannot be observed satisfactorily by separate measurement of the spectra of denatured and native protein. It is more usual and more precise to employ the technique of difference spectrophotometry. For example, a solution of denatured protein can be placed in the reference beam of a double-beam spectrophotometer and a solution of equal concentration of native protein is placed in the sample beam. The resulting difference spectrum (fig. 3.8) shows the perturbations due to differences in molecular environment of the aromatic amino acid residues in native protein as compared with denatured protein. In a similar manner, it is possible to study the conformational changes which are produced by shift of pH, exposure to detergents, urea or salts, or by the binding of a substrate by an enzyme.

A more detailed account of spectrophotometric studies of proteins in the ultraviolet region has been given by Wetlaufer (1962).

3.8. Optical rotatory dispersion of proteins. The optical rotation of a solution as a function of the wavelength at which it is measured is known as the optical rotatory dispersion. The optical rotation of polypeptides is usually quoted as a reduced mean residue rotation, $[R']$, given by the equation:

$$[R'] = \frac{3}{n^2+2}\,\frac{M[\alpha]}{100},$$

where M is the average molecular weight of the amino acids present in the polypeptide, $[\alpha]$ is the specific rotation, and the term, $3/n^2+2$ is a normalizing factor. Theory predicts that the optical rotation is proportional to n^2+2, where n is the refractive index, and the normalizing factor converts the rotation into the value it would be in a solvent of unit refractive index.

For substances which do not absorb in the wavelength region over which the optical rotatory dispersion is measured, a one-term Drude equation:

$$[R'] = a_0\lambda_0^2/(\lambda^2-\lambda_0^2)$$

in which a_0 and λ_0 are constants, relates $[R']$ to the wavelength, λ. This simple equation does not adequately describe the optical rotatory dispersion of native proteins, since the rotatory power for a polypeptide depends partly on the asymmetry of the constituent amino acids and partly on the spatial asymmetry of helical structures. The one-term Drude equation is obeyed, however, by

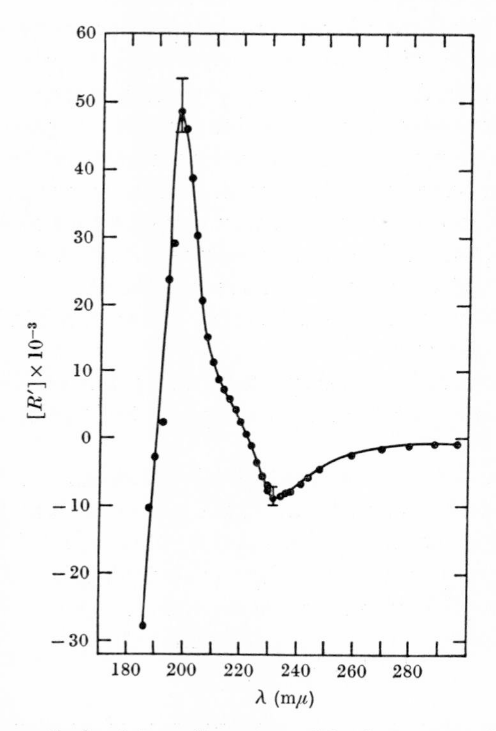

Fig. 3·9. The optical rotatory dispersion of bovine serum albumin in aqueous solution at pH 7·1. [E. R. Blout, I. Schmier & N. S. Simmons (1962). *J. Amer. chem. Soc.* **84**, 3193.]

many proteins when they have a random-coil conformation. Proteins which have a partly or wholly α-helical conformation exhibit anomalous optical rotatory dispersion curves with a trough near 233 mμ (negative Cotton effect) and a maximum near 198 mμ (positive Cotton effect) (fig. 3.9). The relationship between $[R']$ and λ can be expressed in a two-term Drude equation due to Moffitt and Yang:

$$[R'] = a_0\lambda_0^2/(\lambda^2-\lambda_0^2)+b_0\lambda_0^4/(\lambda^2-\lambda_0^2)^2.$$

In this equation, a_0 is strongly dependent on the solvent, while for many polypeptides, which are completely in the α-helical form, λ_0 is approximately 212 mμ and b_0 is approximately $-630°$. If it is assumed that b_0 is zero (i.e. a single-term Drude equation applies) for randomly-coiled polypeptides, it is apparent that the second term in the Moffitt–Yang equation is related to the contribution made to the optical rotation by the helical segments of the polypeptide molecule. By determining the optical rotatory dispersion curve for a polypeptide and computing b_0, the percentage of helix content can be calculated by linear interpolation (table 3.3).

TABLE 3.3. *Percentage of α-helical content of polypeptides in aqueous solution*

	b_0	% Helix from b_0	% Helix from hypo-chromism at 190 mμ	H_1	H_2
Poly-L-glutamic acid	-600	95	—	100*	100*
Paramyosin	-600	95	105	97	95
Myoglobin	-509	78	99	—	-
Myosin	-380	60	—	63	58
Tobacco mosaic virus protein	-160	25	—	—	—
Ribonuclease	-100	16	58	27	26

* Assumed as standard for other determinations.

Determination of the percentage helicity from b_0 may be low if the protein contains segments of left-handed helix as well as of right-handed helix. It should be noted that no account is taken of ordered structures other than the α-helix. Moreover, some proteins show weak Cotton effects in the random coil form. Other factors which may affect optical rotation include side-chain interactions such as the formation of noncovalent bonds, the presence of disulphide bridges, and the presence of prosthetic groups.

Some of these doubtful features in the calculation of the percentage of helical content have been obviated by using a modified two-term Drude equation:

$$[R'] = A_{1(\alpha,\rho)}\lambda_{193}^2/(\lambda^2-\lambda_{193}^2)+A_{2(\alpha,\rho)}\lambda_{225}^2/(\lambda^2-\lambda_{225}^2)$$

in which the subscripts of the λ terms are the wavelengths in mμ. The parameters, $A_{1(\alpha,\rho)}$ and $A_{2(\alpha,\rho)}$ are functions of the α-helical content and random coil content. For polypeptides consisting of only α-helical and random coil segments, these parameters were found to be linearly related:

$$A_{2(\alpha,\rho)} = -0{\cdot}55A_{1(\alpha,\rho)} - 430.$$

Deviation from these lines by a protein indicates that conformations other than the α-helix and random coil are present. If it is assumed that poly-α-L-glutamic acid is completely α-helical at pH 4 ($A_{1(\alpha,\rho)} = +2900$, $A_{2(\alpha,\rho)} = -2050$), and poly-$\alpha$-L-glutamic acid has a completely random conformation at pH 7

$$(A_{1(\alpha,\rho)} = -750,\ A_{2(\alpha,\rho)} = -60),$$

two estimates of the percentage of α-helical content can be made:

$$H_1 = (A_{1(\alpha,\rho)} + 750)/36{\cdot}5,$$
$$H_2 = -(A_{2(\alpha,\rho)} + 60)/19{\cdot}9.$$

When H_1 and H_2 are in agreement (table 3.3), it may be concluded that the molecule contains only the α-helical and random coil structures. When H_1 and H_2 are not in agreement, other conformations are probably present.

A more detailed account of optical rotatory dispersion measurements on proteins and their significance is given by Urnes & Doty (1961). A more general account of the technique and its practice is given by Djerassi (1960).

3.9. X-ray analysis of protein structures. Mention has already been made of the application of X-ray analysis to the determination of the geometry of the peptide bond and the recognition of the right-handed α-helix as the preferred conformation of many simple polypeptides. In this section we shall consider the results which have stemmed from the structural analysis of large proteins in the crystalline state. It might be thought that the determination of the structures of proteins in the crystalline state is an academic exercise for X-ray crystallographers, since proteins do not occur in the crystalline form in living organisms. It might reasonably have been expected that proteins in solution would have a different conformation from that in the solid state and that knowledge of the crystalline structure would throw little light on the relation between protein conformation and biological activity. Several enzymes, however, have been shown to be active in the crystalline state and so it is probable that conformational changes brought about by dissolution in aqueous media are small.

It is impossible in a small textbook to describe the theory and application of X-ray analysis, especially in the protein field, and as Dickerson (1964) has said, 'an oversimplified presentation is little better than no presentation at all'. Accepting the advice of an expert, we shall describe here only the results which X-ray crystallographers have provided for protein chemists and refer the interested reader to Dickerson's review and to the references cited therein for a fuller account of the method. This inevitably means that it is not possible to appreciate here the difficulties encountered in arriving at a complete structure for a protein and the kinds of methods which are used to obviate them. If, however, the reader remembers that Kendrew and Perutz were awarded Nobel prizes for their work on myoglobin and haemoglobin, he will at least accord the work a measure of respect which it is difficult to convey in a treatment as superficial as the present.

Myoglobin is a haem-containing protein, present in vertebrate and invertebrate cells and especially muscle, which combines reversibly with oxygen and serves as an oxygen reservoir. It has a molecular weight of about 18,000 and contains 153 amino acid residues in a single polypeptide chain. Determination of the primary structure by degradative methods and of the secondary and tertiary structures by X-ray analysis were carried out side by side on myoglobin from sperm whale. The primary structure is as follows:

NA1 2 A1 2 3 4 5 6 7 8 9 10 11 12 13 14 15
Val-Leu-Ser-Glu-Gly-Glu-Trp-Glu(NH_2)-Leu-Val-Leu-His-Val-Trp-Ala-Lys-Val-

16 AB1 B1 2 3 4 5 6 7 8 9 10 11 12 13 14
Glu-Ala-Asp-Val-Ala-Gly-His-Gly-Glu(NH_2)-Asp-Ile-Leu-Ile-Arg-Leu-Phe-

15 16 C1 2 3 4 5 6 7 CD1 2 3 4 5 6 7 8
Lys-Ser-His-Pro-Glu-Thr-Leu-Glu-Lys-Phe-Asp-Arg-Phe-Lys-His-Leu-Lys-

D1 2 3 4 5 6 7 E1 2 3 4 5 6 7 8 9 10 11
Thr-Glu-Ala-Glu-Met-Lys-Ala-Ser-Glu-Asp-Leu-Lys-Lys-His-Gly-Val-Thr-Val-

12 13 14 15 16 17 18 19 20 EF1 2 3 4 5 6 7 8 F1
Leu-Thr-Ala-Leu-Gly-Ala-Ile-Leu-Lys-Lys-Lys-Gly-His-His-Glu-Ala-Glu-Leu-

2 3 4 5 6 7 8 9 FG1 2 3 4 5 G1 2 3 4
Lys-Pro-Leu-Ala-Glu(NH_2)-Ser-His-Ala-Thr-Lys-His-Lys-Ile-Pro-Ile-Lys-Tyr-

5 6 7 8 9 10 11 12 13 14 15 16 17 18 19 GH1 2 3
Leu-Glu-Phe-Ile-Ser-Glu-Ala-Ile-Ile-His-Val-Leu-His-Ser-Arg-His-Pro-Gly-

4 5 6 H1 2 3 4 5 6 7 8 9 10
Asp(NH_2)-Phe-Gly-Ala-Asp-Ala-Glu(NH_2)-Gly-Ala-Met-Asp(NH_2)-Lys-Ala-

11 12 13 14 15 16 17 18 19 20 21 22 23 24 HC1 2 3
Leu-Glu-Leu-Phe-Arg-Lys-Asp-Ile-Ala-Ala-Lys-Tyr-Lys-Glu-Leu-Gly-Tyr-

4 5
Glu-(NH_2)-Gly.

A diagrammatic representation of the conformation as revealed by Kendrew's work is given in fig. 3.10. It will be seen that a large proportion of the molecule is α-helical. The non-helical regions differ in composition from one another and, apart from proline, it is not possible to decide which amino acids are most likely to disrupt the formation of a helix. Nearly all the polar groups lie on the surface of the molecule while the interior of the molecule is almost entirely non-polar. The haem group lies in a pocket lined with non-polar groups and there is a large number of van der Waal's contacts between the haem group and neighbouring side-chains of amino acids. The fifth of the six coordination positions of the iron atom is occupied by a ring nitrogen atom of histidine (F8; fig. 3.10). The sixth position is available for the binding of oxygen and this process is probably assisted by another histidine residue (E7; fig. 3.10).

Haemoglobin, which is present in the erythrocytes of vertebrates, is also a haem-containing protein. It combines loosely with oxygen and is responsible for its transport around the body. Its structure is more complex than that of myoglobin, since the molecule contains four chains, two α- and two β-chains, and each carries a haem group. The α- and β-chains have quite similar primary structures and it is likely that they both evolved genetically from a single-chain precursor. There are also resemblances between the primary structures of myoglobin and of the α- and β-chains of haemoglobin. It is perhaps not surprising, therefore, that X-ray analysis has shown that the chains of haemoglobin adopt conformations which resemble that of myoglobin. The four chains of haemoglobin are assembled into a quaternary structure in which there is little contact between like chains, but extensive contacts between α- and β-chains. As with myoglobin, proline residues are present in or near non-helical regions and there is a

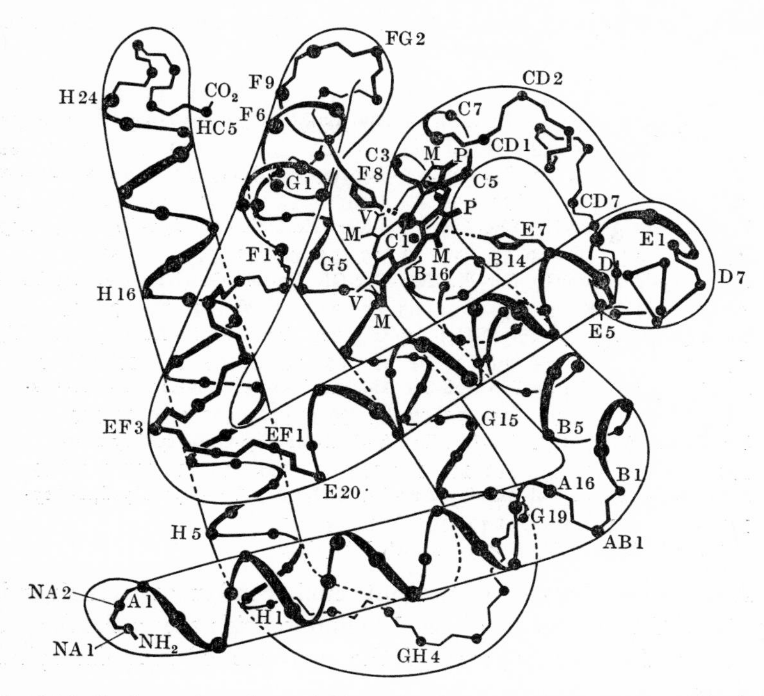

Fig. 3.10. Conformation of myoglobin molecule as revealed by X-ray analysis at a resolution of 2 Å. Large dots represent α-carbon atoms. Stretches of α-helix are represented by a smooth helix with exaggerated perspective and are given labels comprising a single letter and a number. The helical section D 1–D 7 is nearly normal to the paper and is shown by single straight lines connecting α-carbon atoms. Non-helical regions are given labels consisting of two letters and a number. Five-membered rings E 7 and F 8 are imidazole rings which are associated with the haem group. The fainter parallel lines indicate the shape of the molecule as revealed by X-ray analysis at a resolution of 6 Å. [After R. E. Dickerson (1964). In *The Proteins*, vol. 2, p. 634, Ed. by H. Neurath. New York: Academic Press, Inc. and modified to include the *N*-terminal region NA 1–NA 2, which was established by A. B. Edmundson (1965). *Nature*, **205**, 883, as a result of degradative studies on myoglobin.]

short non-helical region at the *N*-terminus. The iron atoms of haemoglobin are approximately tetrahedrally arranged with the closest pair about 25·2 Å apart. Somewhat surprisingly, the two β-chains are closer together in oxyhaemoglobin than they are in reduced haemoglobin.

Lysozyme, an enzyme from hen's egg-white which hydrolyses a mucopolysaccharide component of bacterial cell walls, has also been studied by both degradative and X-ray methods. The sequence of amino acids is as follows:

1 2 3 4 5 6 7 8 9 10 11 12 13 14 15 16 17 18
Lys-Val-Phe-Gly-Arg-CyS-Glu-Leu-Ala-Ala-Ala-Met-Lys-Arg-His-Gly-Leu-Asp-

19 20 21 22 23 24 25 26 27 28 29 30 31 32 33
Asp(NH_2)-Tyr-Arg-Gly-Tyr-Ser-Leu-Gly-Asp(NH_2)-Trp-Val-Cys-Ala-Ala-Lys-

34 35 36 37 38 39 40 41 42 43 44 45
Phe-Glu-Ser-Asp(NH_2)-Phe-Asp(NH_2)-Thr-Glu(NH_2)-Ala-Thr-Asp(NH_2)-Arg-

46 47 48 49 50 51 52 53 54 55 56 57 58 59
Asp(NH_2)-Thr-Asp-Gly-Ser-Thr-Asp-Tyr-Gly-Ile-Leu-Glu(NH_2)-Ile-Asp(NH_2)-

60 61 62 63 64 65 66 67 68 69 70 71 72 73 74
Ser-Arg-Trp-Trp-CyS-Asp(NH_2)-Asp-Gly-Arg-Thr-Pro-Gly-Ser-Arg-Asp(NH_2)-

75 76 77 78 79 80 81 82 83 84 85 86 87 88 89 90
Leu-CyS-Asp(NH_2)-Ile-Pro-CyS-Ser-Ala-Leu-Leu-Ser-Ser-Asp-Ile-Thr-Ala-

91 92 93 94 95 96 97 98 99 100 101 102 103 104 105
Ser-Val-Asp(NH_2)-CyS-Ala-Lys-Lys-Ile-Val-Ser-Asp-Gly-Asp-Gly-Met-

106 107 108 109 110 111 112 113 114 115 116 117 118 119
Asp(NH_2)-Ala-Trp-Val-Ala-Trp-Arg-Asp(NH_2)-Arg-CyS-Lys-Gly-Thr-Asp-

120 121 122 123 124 125 126 127 128 129
Val-Glu(NH_2)-Ala-Trp-Ile-Arg-Gly-CyS-Arg-Leu

A schematic drawing of the main chain conformation is given in fig. 3.11. Only about 55 of the 129 amino acids are in the form of α-helices. Perhaps because of the presence of four disulphide bridges

6 127 30 115 64 80 76 94
(CyS CyS, CyS CyS, CyS CyS and CyS CyS),

the remainder of the molecular conformation is more complex than is found in myoglobin. Some of the amino acids with hydrophobic side chains, Trp^{28}, Trp^{108}, Trp^{111}, Met^{105} and Tyr^{23} are in close proximity and form a kind of hydrophobic box. Some of the hydrophobic groups, however, lie on the surface of the molecule. All of the lysine and arginine residues are external.

In an important and interesting extension of this work, some crystalline lysozyme-inhibitor complexes have been subjected to X-ray analysis. Since the inhibitors behave competitively, they are presumably bound at the active centre of the enzyme. From the low-resolution analysis so far completed, it appears that the following amino acids are in juxtaposition with the inhibitor in the complexes: $Asp(NH_2)^{44}$, $Asp(NH_2)^{46}$-Asp^{48}, Ser^{50}, Asp^{52}, $Glu(NH_2)^{57}$, $Asp(NH_2)^{59}$, Arg^{61}–Trp^{63}, Ser^{72}–Arg^{73}, Lys^{97}, Val^{99}–Asp^{101}, Asp^{103}, Ala^{107}–Ala^{110}, $Asp(NH_2)^{113}$–Arg^{114}.

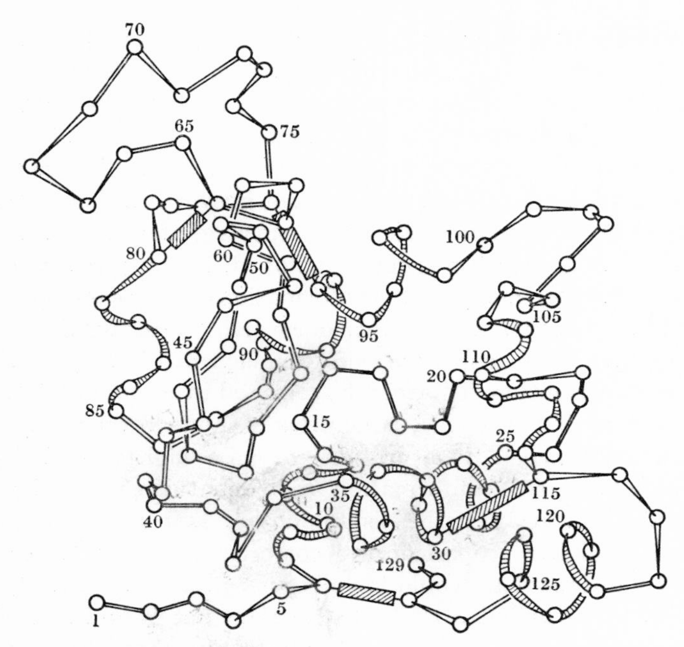

Fig. 3.11. Schematic drawing of the main chain conformation of hen's egg-white lysozyme. [C. C. F. Blake, D. F. Koenig, G. A. Mair, A. C. T. North, D. C. Phillips & V. R. Sarma (1965). *Nature*, **206**, 757.]

3.10. Denaturation of proteins. Exposure of a solution of a protein to heat, extremes of pH, or reagents such as urea, guanidinium chloride or detergents can bring about denaturation. The latter term is unsatisfactory since it has so far defied precise definition. Conformation changes occur during denaturation and these are reflected by changes in, for example, optical rotatory dispersion, titration behaviour and absorption spectra. The precise nature of the conformational change is still debatable, but presumably noncovalent bonds are ruptured to a greater or lesser extent with the consequent exposure to solvent of buried groups and a probable loss of helical character. How much of the helical structure is affected is uncertain and the helical content of a protein does not seem to influence to any appreciable extent the ease of denaturation. It is probable that during denaturation a protein can assume many conformations, separated by small differences in energy level, and these may coexist. Different methods of denaturation may rupture different types of noncovalent bond preferentially. For example, denaturation of a protein by urea is often reversible; removal of the urea by dialysis regenerates native protein with full biological activity. In contrast, while a small increase in temperature may cause reversible denaturation, a large rise in temperature almost always brings about irreversible denaturation. Presumably, if enough weak bonds survive, the remainder will reform if the denaturing agent is removed. If many non-covalent bonds are broken, many more conformations become energetically possible and the statistical chance of reversing denaturation is low. The presence of covalent disulphide bonds probably helps to stabilize the conformation. This may be reflected in a high resistance to denaturation (for example, ribonuclease shows a relatively high stability to thermal denaturation) or in a high probability of reforming the native conformation when denaturing conditions are removed. Conversely, if the disulphide bonds in ribonuclease are reduced, a higher proportion are correctly reformed under oxidizing conditions than would have been expected statistically and so it must be concluded that non-covalent bonds and disulphide bonds act co-operatively in maintaining the native conformation.

There have been many studies of protein denaturation, both theoretical and experimental, but most have considered the phenomenon thermodynamically as an equilibrium process. Treatment of denaturation from a kinetic standpoint, which has rarely been undertaken, may throw more light on the mechanism of this complex process.

4. Chemical synthesis of peptides

4.1. Basic principles and strategy. The synthesis of a dipeptide, $NH_3^{\oplus}$.CHR.CO.NH.CHR′.CO.$O^{\ominus}$, from the constituent amino acids involves forming the peptide bond —CO.NH— so that the sequence of amino acids is correct and racemization does not occur at the asymmetric α-carbon atoms. The latter point does not arise, of course, with glycine.

In the synthesis of higher peptides, the polypeptide chain can be built up one unit at a time in either direction. For example, an octapeptide could be synthesized in several stages proceeding through a dipeptide, tripeptide, tetrapeptide and so on. Alternatively, the octapeptide could be built up from two tetrapeptide units, which in turn might be either built up one unit at a time or formed from two dipeptides. Apart from chemical considerations, the overall yield would depend on the route selected. For example, if the synthesis of each peptide bond could be achieved with a yield of 80 per cent, the stepwise procedure of adding one amino acid residue at a time would give an overall yield of 21 per cent based on the first two amino acids used. The stepwise synthesis of two tetrapeptides followed by the union of these would afford an overall yield of 41 per cent based on the amino acids used in the syntheses of the first two dipeptides. Finally, the synthesis which proceeds through four dipeptides to two tetrapeptides and thence to the octapeptide would give an overall yield of 51 per cent. Other factors must also be borne in mind. For example, the risk of racemization varies with the synthetic route selected (§ **4.6**). In addition, the prices of the cheapest and most expensive amino acids may differ by a factor of 100. It is advantageous to have glycine as the *C*-terminal residue of a large unit whose carboxyl group is to be involved in the synthesis of a peptide bond, since the risk of racemization is eliminated (§ **4.6**). Further, it is cheaper to postpone, if possible, the introduction of the more expensive amino acids until the later stages of the synthesis.

The synthesis of even a small peptide thus requires considerable forethought.

The synthesis of a dipeptide, in general, involves four steps: (*a*) protection of the amino group of the amino acid which is to be the *N*-terminal residue in the dipeptide (LVI → LVII); (*b*) protection of the carboxyl group of the amino acid which is to be the *C*-terminal residue in the dipeptide (LVIII → LIX); (*c*) conversion of the carboxyl group of the *N*-protected amino acid into a suitable derivative (LVII → LX) which can then undergo nucleophilic attack by the amino group of the *C*-protected amino acid to give a fully protected dipeptide (LIX + LX → LXI); (*d*) removal of protecting groups [LXI → LXII (or LXIII) → LXIV].

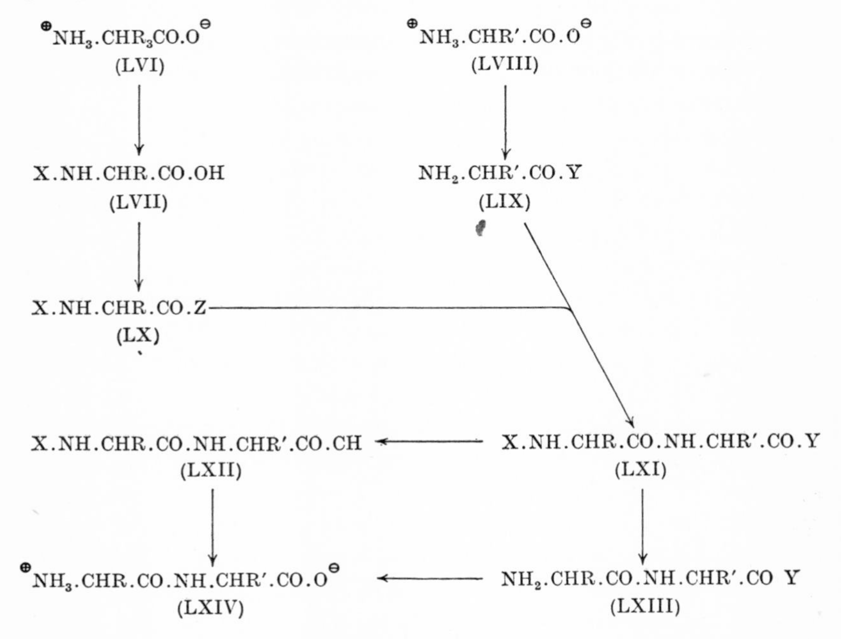

The formation of the peptide bond (LIX + LX → LXI) requires the nitrogen atom in component (LIX) to be unprotonated so that it is nucleophilic. The protecting group, Y, lowers the pK_a of the $—NH_3^{\oplus}$ group and facilitates removal of the proton. Protected derivatives (LIX) are also more soluble than free amino acids in organic solvents which are the usual media for the condensation step. Similarly, the derivatives (LVII) are more soluble than amino acids in organic solvents. The carbonyl-carbon atom in the derivative (LVII) is more positive than the corresponding carbon atom in the dipolar ion (LVI). The positive character of this carbon atom is further increased by forming, for example, the corresponding acid chloride (LX; Z = Cl), azide (LX; $Z = N_3$) or *p*-nitrophenyl ester (LX; $Z = p\text{-}NO_2.C_6H_4.O$).

Clearly, a similar sequence of reactions can be carried out starting with an *N*-protected peptide (e.g. LXII instead of LVII) and a *C*-protected peptide (e.g. LXIII instead of LIX) to give ultimately longer peptides. Only when the required number of peptide bonds had been formed would the product be completely stripped of protecting groups. Although Emil Fischer proposed this general route for the synthesis of peptides at the beginning of the century, the lack of suitable reagents for protecting amino groups virtually precluded its use for the next thirty years. Indeed, Fischer was driven to synthesize peptides by the following route:

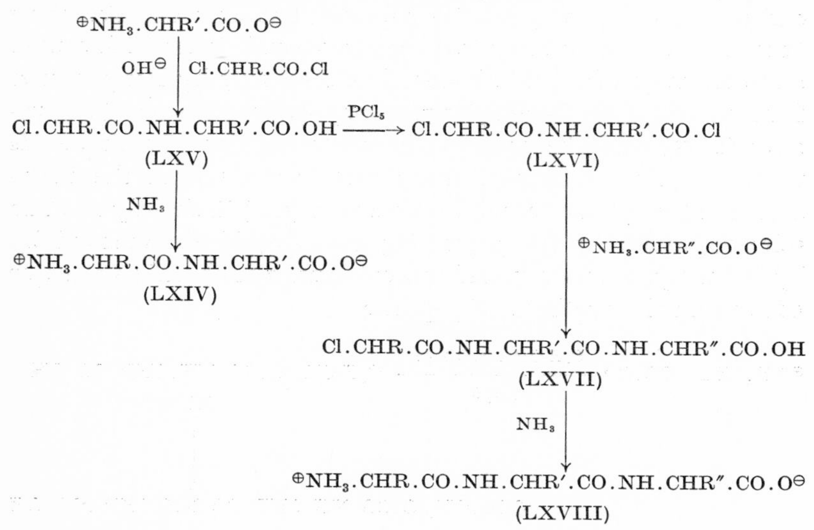

Treatment of an amino acid in aqueous solution with an α-chloroacyl chloride in the presence of base gave the α-chloroacyl derivative (LXV). This could either be converted into a dipeptide by treatment with ammonia (LXV → LXIV) or treated with phosphorus pentachloride to give the acid chloride (LXVI). The latter could be coupled with another amino acid or even a peptide in order to build up longer peptides. The α-chloroacyl group serves to protect the amino group and need not be removed, since it is itself transformed into the *N*-terminal residue at the end of the synthesis. Although Fischer used this route to synthesize an octadecapeptide, it suffers from numerous disadvantages. Thus, α-chloroacyl halides are not easily accessible in stereochemically pure form, use of acyl chlorides to form peptide bonds is attended by considerable risk of racemization, and amination of *N*-α-chloroacetylamino acids can give rise to a variety of side-reactions depending on the nature of the derivative (LXV). These difficulties need not concern us further here, for despite the historical interest attaching to Fischer's work, subsequent work has made this route virtually obsolete.

4.2. Protection of amino groups. *N*-Benzoyl, *N*-ethoxycarbonyl and *N*-toluene-*p*-sulphonyl groups were used in the earliest attempts to synthesize peptides, but were unsatisfactory because their removal required the use of conditions which caused some cleavage of peptide bonds. The discovery by Bergmann & Zervas (1932) that *N*-benzyloxycarbonyl ('carbobenzoxy') derivatives (LXIX) could be prepared from the reaction of amino acids with benzyl chloroformate (itself readily made from the interaction of benzyl alcohol with excess phosgene) in slightly alkaline solution, and that the protecting group could be removed by hydrogenolysis over a palladium catalyst, was therefore a major advance in the synthesis of peptides.

$$^{\oplus}NH_3.CHR.CO.O^{\ominus} \xrightarrow[OH^{\ominus}]{C_6H_5CH_2.O.CO.Cl} \underset{\text{(LXIX)}}{C_6H_5CH_2.O.CO.NH.CHR.CO.OH}$$

$$\downarrow$$

$$C_6H_5.CH_2.O.CO.NH.CHR.CO.NH.CHR'.CO.OH$$

$$\downarrow H_2/Pd$$

$$^{\oplus}NH_3.CHR.CO.NH.CHR.CO.O^{\ominus}$$

The products of hydrogenolysis, toluene and carbon dioxide, are inert and easily removed. Peptide bonds, of course, are not affected by hydrogenolysis.

The *N*-benzyloxycarbonyl group can also be removed by reduction with sodium in liquid ammonia. This method is less selective than hydrogenolysis, however, since other groups such as *S*-benzyl and *N*-toluene-*p*-sulphonyl are also cleaved. The *N*-benzyloxycarbonyl group can be conveniently removed by a solution of dry hydrogen bromide in warm acetic acid or nitromethane. The mechanism of this reaction probably involves an initial protonation of the nitrogen atom, since this appears to be more nucleophilic than the carbonyl-oxygen atom in urethanes. Protonation promotes a subsequent nucleophilic attack by bromide ion on the benzyl group.

$$C_6H_5.CH_2.O.CO.NH.CHR.CO \ldots \xrightarrow{HBr} [C_6H_5.CH_2.O.CO.\overset{\oplus}{N}H(H).CHR.CO \ldots] \; Br^{\ominus}$$

$$\downarrow$$

$$C_6H_5.CH_2.Br + CO_2 + NH_2.CHR.CO \ldots \xrightarrow{H^{\oplus}} \overset{\oplus}{N}H_3.CHR.CO \ldots$$

The related *N*-t-butyloxycarbonyl group, which is best introduced using t-butyl azidoformate (LXX) since the chloroformate is very unstable, differs from the *N*-benzyloxycarbonyl group in its stability towards hydrogenolysis and sodium in liquid ammonia and its extreme lability towards acids; cold trifluoroacetic acid or a solution of dry hydrogen chloride in ether remove the *N*-t-butyloxycarbonyl group in a few minutes but do not affect *N*-benzyloxycarbonyl derivatives.

$$\underset{\text{t-Butyl carbazate}}{NH_2NH.CO.O.C(CH_3)_3} \xrightarrow{HNO_2} \underset{(LXX)}{N_3CO.O.C(CH_3)_3}$$

$$\downarrow \; OH^{\ominus} \;|\; \overset{\oplus}{N}H_3.CHR.CO.O^{\ominus}$$

$$(CH_3)_3C.O.CO.NH.CHR.CO.OH$$

$$\downarrow \; HCl$$

$$(CH_3)_3C.Cl + CO_2 + \overset{\oplus}{N}H_3.CHR.CO.OH \quad Cl^{\ominus}$$

The *N*-benzyl- and *N*-t-butyl-oxycarbonyl groups are probably the most useful means of protecting amino groups at present.

The toluene-*p*-sulphonyl ('tosyl') group was used for blocking α-amino groups at an early stage in the development of methods of peptide synthesis, but its full potential was not realized until it was discovered that it could be removed by the reducing action of sodium in liquid ammonia. More recently, it has proved useful for protecting the ϵ-amino group of lysine and the guanidino group of arginine.

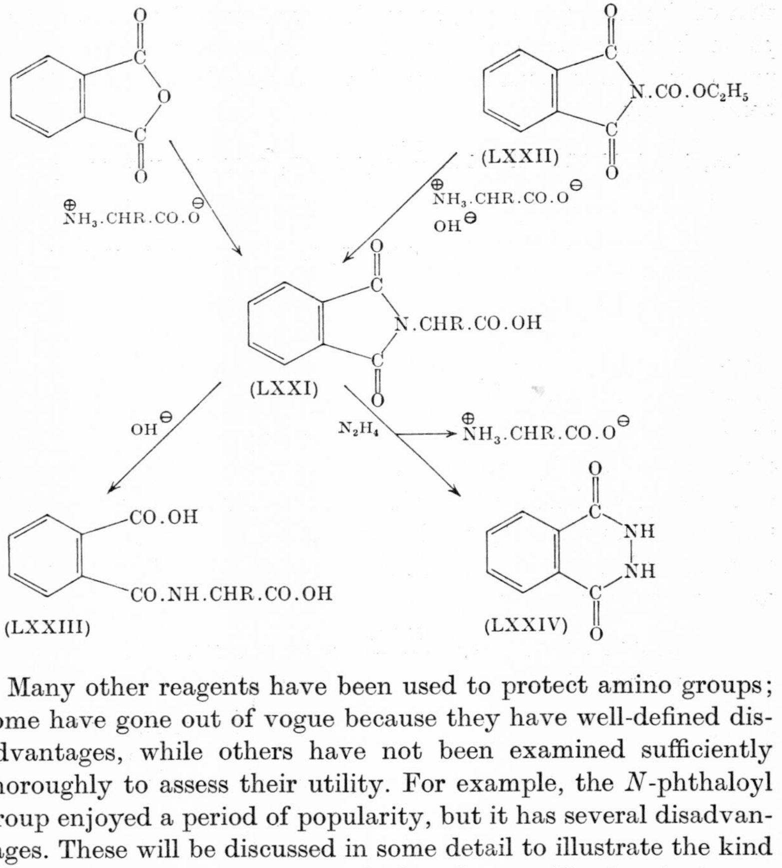

$N.CO.OC_2H_5$ (LXXII)

$\overset{\oplus}{N}H_3.CHR.CO.O^{\ominus}$ $\quad$ $\overset{\oplus}{N}H_3.CHR.CO.O^{\ominus}$, $OH^{\ominus}$

$N.CHR.CO.OH$ (LXXI)

$OH^{\ominus}$ $\quad$ N_2H_4 $\longrightarrow$ $\overset{\oplus}{N}H_3.CHR.CO.O^{\ominus}$

$CO.OH$, $CO.NH.CHR.CO.OH$ (LXXIII) $\quad$ NH, NH (LXXIV)

Many other reagents have been used to protect amino groups; some have gone out of vogue because they have well-defined disadvantages, while others have not been examined sufficiently thoroughly to assess their utility. For example, the *N*-phthaloyl group enjoyed a period of popularity, but it has several disadvantages. These will be discussed in some detail to illustrate the kind of screening which is necessary before the peptide chemist is prepared to accept a reagent as a standard weapon in his armoury. Preparation of *N*-phthaloylamino acids (LXXI) by the direct interaction of phthalic anhydride and amino acids in the fused state or in organic solvents frequently leads to partial or complete racemization. This difficulty can be overcome, however, by forming the *N*-phthaloylamino acids from *N*-ethoxycarbonyl-phthalimide (LXXII) and amino acids in aqueous alkaline solution.

Again, the *N*-phthaloyl group is sensitive to alkaline conditions such as might be used, for example, to hydrolyse an ester of an *N*-phthaloyl-peptide to the free acid. The five-membered ring is opened and derivatives of phthalamic acid (LXXIII) are formed. The *N*-phthalamoyl group cannot be removed under mild conditions. The usual method for removing *N*-phthaloyl groups involves treatment with hydrazine hydrate under reflux. Sparingly soluble tetrahydro-phthalazine-1,4-dione (LXXIV) is formed and can be separated easily from the free peptide. Attempted cleavage of the *N*-phthaloyl group from esters of *N*-phthaloyl-peptides usually leads to the formation of peptide hydrazides unless the carbonyl group is protected with a t-butyl ester group.

Another reagent for protecting amino groups, whose potential value is still being assessed, is *o*-nitrophenylsulphenyl chloride (LXXV). The reagent readily acylates amino acids in aqueous solution:

$$o\text{-}NO_2C_6H_4\text{—}S.Cl \;+\; \overset{\oplus}{N}H_3.CHR.CO.O^{\ominus} \xrightarrow{OH^{\ominus}} o\text{-}NO_2C_6H_4\text{—}S.NH.CHR.CO.OH$$

(LXXV) (LXXVI)

The resulting *o*-nitrophenylsulphenylamino acids (LXXVI) are usually isolated and stored in the form of their dicyclohexylammonium salts. The *o*-nitrophenylsulphenyl group is extremely sensitive to acids and can be removed, for example, by a solution of hydrogen chloride in ether in a few minutes at room temperature.

4.3. Protection of carboxyl groups. There are numerous instances in the literature where the carboxyl group of the amino acid, which eventually forms the *C*-terminal residue of a peptide, has not been protected. In such cases, formation of the peptide bond has to be carried out in aqueous alkaline solution, aqueous because amino acids are not appreciably soluble in organic solvents and alkaline because a proton must be dissociated from the $\text{—}\overset{\oplus}{N}H_3$

group of the amino acid to make it nucleophilic and therefore capable of reacting with the component (LX). On the face of it, the absence of a protecting group at the *C*-terminus is attractive because it saves a step later in the synthesis. This approach, however, is by no means universally possible. Depending on the ease of solvolysis of the component (LX), formation of the peptide bond may have to be conducted in an aprotic solvent. In addition, the use of alkaline conditions always increases the danger of racemization.

In general, carboxyl groups of amino acids are esterified before the peptide bond is formed. Amino acids differ profoundly in properties from their esters. The former are dipolar ions while the esters are simple bases which are soluble in aprotic solvents. There is also a striking difference in the pK_a of the $—\overset{\oplus}{N}H_3$. Since the carboxylic anion, $—CO.O^{\ominus}$, is electron-donating while $—CO.OR$ is electron-withdrawing, it is easier to detach a proton from $\overset{\oplus}{N}H_3.CHR'.CO.OR$ than from $\overset{\oplus}{N}H_3.CHR'.CO.O^{\ominus}$. Esters of amino acids can be liberated from their hydrochlorides by shaking with a solution of ammonia in chloroform before peptide synthesis or *in situ* during peptide synthesis by the addition of a tertiary base. The former method is preferable because the presence of tertiary base can sometimes lead to racemization.

Alkyl esters are frequently used in masking carboxyl groups in peptide synthesis. The methyl or ethyl ester hydrochlorides are readily obtained by passing dry hydrogen chloride into a suspension of the amino acid in dry methanol or ethanol. Amino acids can also be conveniently esterified with a mixture of thionyl chloride and an alcohol.

For the removal of the ester group after peptide synthesis, alkaline hydrolysis is most commonly used, but a number of difficulties have been found. As mentioned above, alkaline conditions may encourage racemization. In addition, some groups used for protecting amino groups such as *N*-phthaloyl and *N*-trifluoroacetyl are labile to alkaline conditions. Alkyl esters are particularly useful, however, if the peptide derivative is to be extended at the *C*-terminus by the azide route, since the intermediate hydrazide is readily obtained from the ester by treatment with hydrazine.

Benzyl esters are frequently more convenient than methyl or ethyl esters, since they can be reductively cleaved to the free acid and toluene by catalytic hydrogenolysis (cf. the cleavage of *N*-benzyloxy carbonyl groups by hydrogenolysis). Treatment with hydrogen bromide in acetic acid also removes the benzyl ester group, although this reaction is less rapid than the cleavage of the *N*-benzyloxycarbonyl group. In suitable cases, the latter process can be achieved selectively leaving a benzyl ester of a peptide, a convenient derivative for extension of the peptide chain at the *N*-terminus. For this particular approach, *p*-nitrobenzyl esters offer advantages, since they are more resistant than benzyl esters to hydrogen bromide in acetic acid but they are readily cleaved by hydrogenolysis.

In contrast, t-butyl esters, like t-butyloxycarbonyl derivatives, are resistant to hydrogenolysis but are very sensitive to anhydrous acids. Dry hydrogen chloride in a suitable solvent or anhydrous trifluoroacetic acid are effective and mild reagents for this purpose. t-Butyl esters are formed when amino acids or their *N*-benzyloxycarbonyl derivatives are allowed to react with isobutylene in the presence of concentrated sulphuric acid. Alternatively, they can be made by a transesterification reaction using t-butyl acetate and perchloric acid as catalyst. If *N*-benzyloxycarbonyl derivatives are obtained, the *N*-protecting group can be selectively removed by hydrogenolysis.

4.4. Protection of side-chains of amino acids. The ϵ-amino group in the side-chain of lysine can be blocked by any of the methods described for α-amino groups. It is often necessary to employ separate protecting groups of differing stability for the α- and ϵ-amino groups. The protecting group on the α-position can then be selectively removed to permit the coupling of further amino acid residues. The protecting group on the ϵ-position is left intact until the end of the synthesis. The synthesis of lysine derivatives with different protecting groups on the α- and ϵ-positions presents little difficulty. Lysine chelates with the cupric ion by means of the α-amino group and the carboxylic anion (LXXVII); the ϵ-amino group can thus be selectively protected and decomposition of the copper chelate with hydrogen sulphide liberates the lysine derivative (e.g. LXXVIII). The α-amino group can then be blocked with a different reagent. For example, the ϵ-amino group can first be masked with the *N*-t-butyloxycarbonyl group and the α-amino group can then be blocked with the *N*-benzyloxycarbonyl group (LXXIX). After forming a peptide bond involving the carbonyl group of lysine and the amino group of a t-butyl ester of an amino acid (LXXX), the *N*-benzyloxycarbonyl group can be removed by hydrogenolysis (LXXXI). The liberated amino group can be coupled with the *N*-t-butyloxycarbonyl derivative of an amino acid to give a fully protected tripeptide (LXXXII). Finally, all remaining protecting groups can be removed with anhydrous hydrogen chloride or trifluoroacetic acid to give the free tripeptide (LXXXIII).

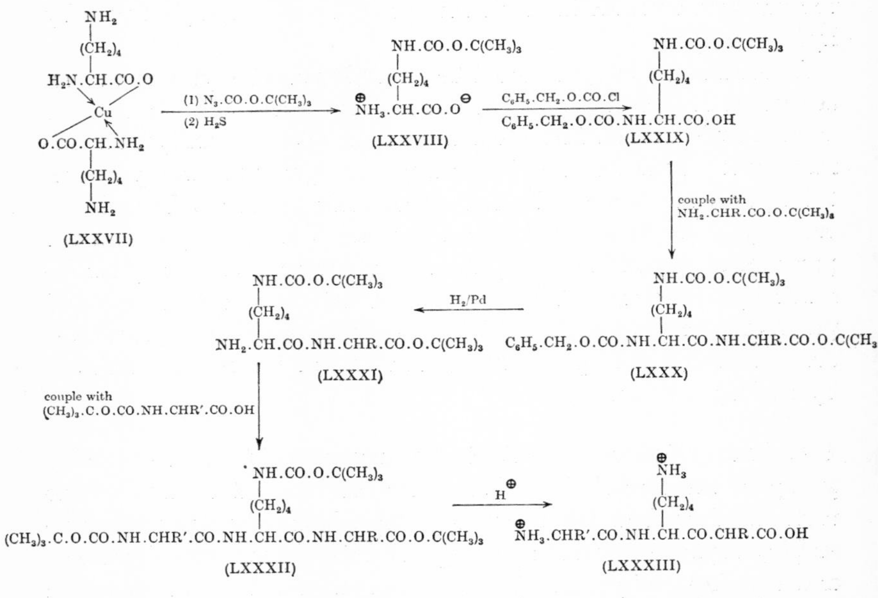

The β-carboxyl group of aspartic acid and the γ-carbonyl group of glutamic acid must be masked during the formation of a peptide bond involving the α-carboxyl group of these amino acids. Benzyl and t-butyl esters are preferred to methyl and ethyl esters for the reasons mentioned above, unless it is desired to convert the aspartyl to asparaginyl and glutamyl to glutaminyl residues. The final choice will depend on whether or not the peptide chain is to be extended at the *N*-terminus. If so, it is necessary to choose a protecting group for the *N*-terminus which can be removed without affecting the ester groups in the side chains of aspartic or glutamic acids. Fortunately, the carboxyl groups in the side-chains of aspartic and glutamic acids are esterified preferentially and it is possible to obtain the desired esters by direct esterification or transesterification under mild conditions.

The thiol group in the side-chain of cysteine is a powerful nucleophile, especially in the anionic form, and it must therefore be protected during peptide synthesis. The *S*-benzyl group was the first to be tried and is probably still the most commonly used, although several other masking groups have been examined. *S*-Benzylcysteine (LXXXIV) is prepared by dissolving cysteine in liquid ammonia, adding sufficient sodium to convert the thiol group into the anion, and then treating with benzyl chloride. The benzyl group is removed by reduction with excess sodium in liquid ammonia. The imidazole ring of histidine is also nucleophilic and can be protected by the *N*-benzyl group (LXXXV) in a similar manner to that used with cysteine.

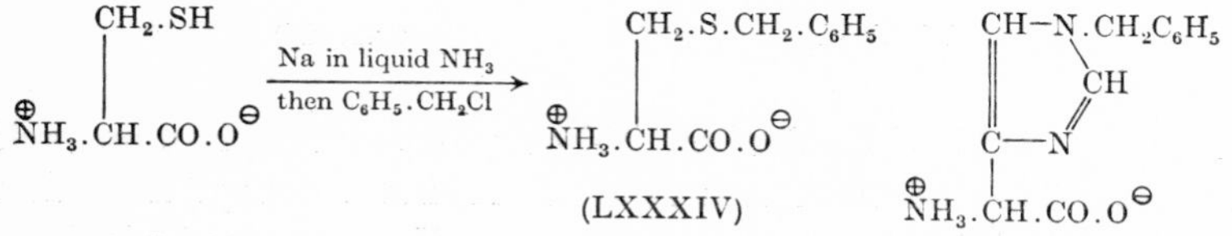

$$\overset{\oplus}{N}H_3.\underset{}{CH}(CH_2.SH).CO.O^{\ominus} \xrightarrow[\text{then } C_6H_5.CH_2Cl]{\text{Na in liquid } NH_3} \overset{\oplus}{N}H_3.CH(CH_2.S.CH_2.C_6H_5).CO.O^{\ominus}$$

(LXXXIV)

CH—N.CH$_2$C$_6$H$_5$ / CH / C—N (imidazole ring), $\overset{\oplus}{N}H_3.CH.CO.O^{\ominus}$

(LXXXV)

The guanidino group in the side-chain of arginine is strongly basic (pK_a ~ 13) and occasionally the proton alone is sufficient to protect it from entering into side reactions during peptide synthesis. Methods are available, however, for masking the guanidino group with an *N*-toluene-*p*-sulphonyl or *N*-benzyloxycarbonyl group and the removal of these at the end of peptide synthesis is accomplished by the same methods that are used to unmask amino groups. Alternatively, arginine can be nitrated to *N*-nitroarginine and the presence of the strongly electron-withdrawing nitro group almost eliminates the basic properties and nucleophilicity of the guanidino group. The *N*-nitro group can be removed by hydrogenolysis, although the reaction is usually slow and requires a high concentration of catalyst.

More detailed accounts of methods used for introducing and removing protecting groups have been given by Goodman &

Kenner (1957), Greenstein & Winitz (1961), Hofmann & Katsoyannis (1963), and Schröder & Lübke (1965).

4.5. Formation of peptide bonds. Most methods of forming a peptide bond involve the acylation of the amino group of one amino acid derivative by some suitable, reactive derivative of an *N*-acylamino acid. The obvious method of converting the carboxyl group into the acid chloride, while popular in the early days of peptide chemistry, is unsatisfactory because side-reactions occur. For example, the acid chloride (LXXXVI) may spontaneously isomerize to an oxazolonium chloride (LXXXVII) and the latter readily racemizes. This is especially important in the case of *N*-acylpeptide acid chlorides.

```
   NH———CH.R'                      ⊕  Cl⊖
   |     |                         NH———CH.R'
 R.C     CO.Cl    ——————→          ||    |
    \\                           R.C     CO
     O                              \   /
                                      O

  (LXXXVI)                         (LXXXVII)
```

Secondly, acid chlorides of *N*-benzyloxycarbonylamino acids (LXXXVIII) are unstable and tend to eliminate benzyl chloride to give oxazolid-2,5-diones (also known as *N*-carboxyanhydrides or Leuchs' anhydrides) (LXXXIX). The latter, which may be regarded as cyclic unsymmetrical anhydrides, polymerize readily in the presence of bases as initiators to give polyamino acids (XC). Oxazolid-2,5-diones are more commonly obtained by the interaction of an amino acid and phosgene.

```
                CO———NH                              CO———NH
                |     |                              |     |
 C6H5.CH2.O     CH.R   ——————→  C6H5.CH2Cl  +        O     CH.R   (LXXXIX)
          Cl.C                                ↗        \   /
             ||                              /           C
             O                              /            ||
                                           /             O
   (LXXXVIII)                                            |
                 ⊕               ⊖                       | R'.NH2
                 NH3.CHR.CO.O  +  COCl2                   ↓
                                           H.[NH.CHR.CO]n.NHR' + nCO2

                                                   (XC)
```

Acyl azides (e.g. XCI) have been used as reagents for synthesizing peptides since the time that the peptide bond was first identified in proteins. The synthesis is accomplished without the risk of racemization and this explains the continued popularity of the method. There are, however, some disadvantages; one of these is the number of steps involved:

$$C_6H_5.CH_2.O.CO.NH.CHR.CO.OH \longrightarrow C_6H_5.CH_2.O.CO.NH.CHR.CO.OCH_3$$

$$\downarrow N_2H_4$$

$$\underset{\text{(XCI)}}{C_6H_5.CH_2.O.CO.NH.CHR.CO.N_3} \xleftarrow{HNO_2} C_6H_5.CH_2.O.CO.NH.CHR.CO.NH.NH_2$$

$$\downarrow NH_2.CHR'.CO.OCH_3$$

$$\underset{\text{(XCII)}}{C_6H_5.CH_2.O.CO.NH.CHR.CO.NH.CHR'.CO.OCH_3}.$$

Clearly, the protected peptide (XCII) can be converted into the corresponding azide for extending the chain from the *C*-terminus.

The possibility that the acyl azide may undergo the Curtius rearrangement to isocyanate (XCIII) is a further disadvantage of the use of acyl azides for synthesizing peptides. If this occurs in the presence of an ester of an amino acid or peptide, a urea (XCIV) will be formed:

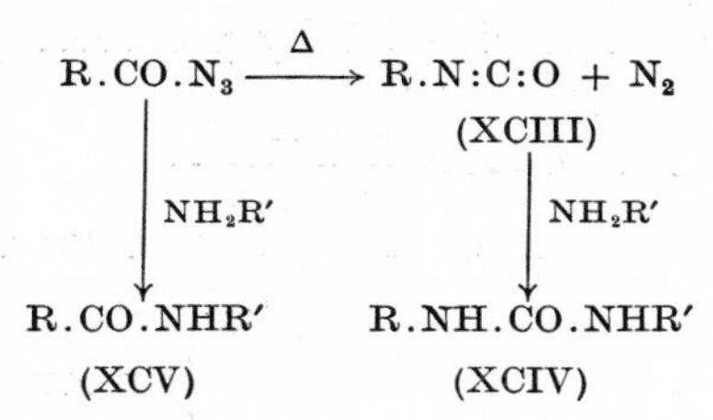

Even if this side-reaction occurs to only a small extent, the desired product (XCV) may be difficult to purify. It is most important, therefore, to keep the temperature low, especially since acyl azides often couple rather slowly with amino esters. Since the *N*-phthaloyl group is sensitive to hydrazine, it is best not to use it as a protecting group for the amino acid which is to be converted into an azide derivative.

Acid anhydrides constitute another type of acylating agent used for synthesizing peptides. Obviously, it would be inefficient to use a symmetrical anhydride of an *N*-acylamino acid, since one molecule of free *N*-acylamino acid would be formed for each molecule of peptide synthesized. A more suitable reagent is an unsymmetrical acid anhydride (XCVI). A nucleophile, such as an amino compound, may attack either of the carbonyl-carbon atoms in an unsymmetrical acid anhydride. Only one of these possibilities leads to the desired product.

$$\underset{\text{(XCVI)}}{\mathrm{R.CO.O.CO.R'}} + \mathrm{R''NH_2} \begin{cases} \nearrow \mathrm{R.CO.HNR''} + \mathrm{R'.CO.OH} \\ \searrow \mathrm{R'.CO.HNR''} + \mathrm{R.CO.OH} \end{cases}$$

Much thought and ingenuity has been devoted to the search for suitable acid anhydrides. Instability of one of the products of nucleophilic attack, steric and electronic factors have all been invoked to direct the reaction in the desired direction. For example, a solution of an *N*-acylamino acid in an inert organic solvent containing one mole of a tertiary base can be treated with alkyl chloroformate at low temperatures to give an unsymmetrical anhydride (XCVII) which is a derivative of carbonic acid. Without isolation, the anhydride is allowed to react with an ester of an amino acid or peptide to give a fully protected peptide (XCVIII).

$$\mathrm{C_6H_5.CH_2.O.CO.NH.CHR'.CO.OH} \xrightarrow[\mathrm{(C_2H_5)_3N}]{\mathrm{RO.CO.Cl}}$$

$$\underset{\text{(XCVII)}}{\mathrm{C_6H_5.CH_2.O.CO.NH.CHR'.CO.O.CO.OR}} + \mathrm{(C_2H_5)_3\overset{\oplus}{N}HCl^{\ominus}}$$

$$\Big\downarrow \mathrm{NH_2.CHR''.CO.OC_2H_5}$$

$$\mathrm{CO_2} + \mathrm{R.OH} + \underset{\text{(XCVIII)}}{\mathrm{C_6H_5.CH_2.O.CO.NH.CHR'.CO.NH.CHR''.CO.OC_2H_5}}$$

In this case, the instability of the half-ester of carbonic acid thermodynamically favours the desired reaction. In addition, the carbonyl-carbon atom in the carbonic acid moiety of the anhydride is less electrophilic than the other carbonyl-carbon atom. In spite of these two favourable factors, it must be emphasized that the choice of solvent is important. Not only must it be aprotic for satisfactory formation of the anhydride (XCVII), but solvation of the latter can profoundly influence the direction of its cleavage by a nucleophile. Nevertheless, this method has been widely used in the last fifteen years. An example of the influence of steric factors in controlling the direction of cleavage of an unsymmetrical acid anhydride is provided by the use of intermediates which are anhydrides of diphenylacetic acid (XCIX). The latter are obtained when *N*-acylamino acids react with diphenylketen in the presence of tertiary base. The bulky phenyl groups direct nucleophilic attack to the carbonyl-carbon atom derived from the *N*-acylamino acid giving a protected peptide and diphenylacetic acid.

$$C_6H_5.CH_2.O.CO.NH.CHR.CO.OH \xrightarrow[(C_2H_5)_3N]{(C_6H_5)_2C:C:O} C_6H_5.CH_2.O.CO.NH.CHR.CO.O.CO.CH(C_6H_5)_2 \quad \text{(XCIX)}$$

$$\downarrow NH_2.CHR'.CO.OC_2H_5$$

$$(C_6H_5)_2CH.CO.OH + C_6H_5.CH_2.O.CO.NH.CHR.CO.NH.CHR'.CO.OC_2H_5$$

It is sometimes possible to couple intermediate unsymmetrical acid anhydrides (e.g. those derived from diphenylacetic acid) with a free amino acid in aqueous alkaline solution instead of with an ester in an organic solvent, but this is not usual, in spite of its apparent advantages (see p. 80), because of the danger of hydrolysing the anhydride.

The unsymmetrical acid anhydrides used in the synthesis of peptides need not be derived from carboxylic acids. For example, treatment of an *N*-acylamino acid with tetraethyl pyrophosphite gives an anhydride (C), which reacts with esters of amino acids to give protected peptides.

$$C_6H_5.CH_2.O.CO.NH.CHR.CO.OH + (C_2H_5O)_2P.O.P(OC_2H_5)_2 \longrightarrow C_6H_5.CH_2.O.CO.NH.CHR.CO.O.P(OC_2H_5)_2 \quad \text{(C)}$$

$$\downarrow NH_2.CHR'.CO.OC_2H_5$$

$$C_6H_5.CH_2.O.CO.NH.CHR.CO.NH.CHR'.CO.OC_2H_5$$

A widely used method of forming peptide bonds involves the use of 'active esters'. This nomenclature is unsatisfactory because, most esters are reactive to a degree. Also the term, 'active esters', carries the possible implication that the compounds are labelled with a radioactive isotope. Unfortunately, no suitable alternative terminology has been suggested and the existing name will be retained here. In one sense, an ester may be regarded as an unsymmetrical acid anhydride which is derived from a carboxylic acid and an exceedingly weak acid such as methanol ($pK_a = 15{\cdot}5$). The analogy is obviously rather tenuous in the case of alkyl esters, but, for example, *p*-nitrophenyl esters (pK_a of *p*-nitrophenol = 7·15) fall in that part of the spectrum of reactivity between acid anhydrides and alkyl esters. Since active esters can be cleaved by a nucleophile in one direction only, they offer at least one advantage over unsymmetrical acid anhydrides. A second attribute is their stability; for example, *p*-nitrophenyl esters of *N*-acylamino acids can usually be crystallized and stored indefinitely.

The reactivity of an ester, R.CO.OR′, depends on the withdrawal of electrons by R and R′ so that the carbonyl-carbon atom is made more prone to nucleophilic attack. If electron-withdrawal depends on inductive effects, R exerts more influence than R′, since the latter is separated from the carbonyl group by an oxygen atom. Cyanomethyl esters (CI), formed by the reaction of chloroacetonitrile with *N*-acylamino acids in the presence of tertiary base, owe their reactivity to the inductive withdrawal of electrons by the —C⋮N group. They react with esters of amino acids rather slowly to give peptide derivatives.

$$C_6H_5.CH_2.O.CO.NH.CHR.CO.OH \xrightarrow[(C_2H_5)_3N]{ClCH_2.CN} C_6H_5.CH_2.O.CO.NH.CHR.CO.O.CH_2.C{\vdots}N \quad (CI)$$

$$\downarrow NH_2.CHR'.CO.OC_2H_5$$

$$C_6H_5.CH_2.O.CO.NH.CHR.CO.NH.CHR'.CO.OC_2H_5$$

In the case of aryl esters, resonance stabilizes the phenoxide anion and makes it a good leaving group. (Similarly, the resonance stabilization of the carboxylate anion partly accounts for the reactivity of acid anhydrides). This effect, coupled with electron-withdrawal from the carbonyl group, accentuates the reactivity of aryl esters, and *p*-nitrophenyl esters of *N*-acylamino acids (CII) have found wide application for the synthesis of peptides.

$$(CH_3)_3.C.O.CO.NH.CHR.CO.O.C_6H_4.NO_2 \xrightarrow{NH_2.CHR'.CO.O.C(CH_3)_3} (CH_3)_3.C.O.CO.NH.CHR.CO.NH.CHR'.CO.O.C(CH_3)_3$$

(CII)

Aryl esters (CII) are obtained from the condensation of the acid and phenol in the presence of *NN'*-dicyclohexylcarbodi-imide (see p. 91) or by transesterification of an *N*-acylamino acid using a diaryl sulphite (CIII) or triaryl phosphite. Incidentally, these last two reagents are good examples of active esters whose reactivity partly stems from the stability of the departing anion.

$$R.CO.OH + (NO_2.C_6H_4.O)_2SO \xrightarrow{\text{pyridine}} R.CO.O.C_6H_4.NO_2 + NO_2.C_6H_4.O.S(=O)O^{\ominus}$$

(CIII)

An unusual method of peptide synthesis uses isoxazolium salts (e.g. CIV) as starting materials and activated enol esters of the type (CVIII) are formed as intermediates, probably by the following mechanism:

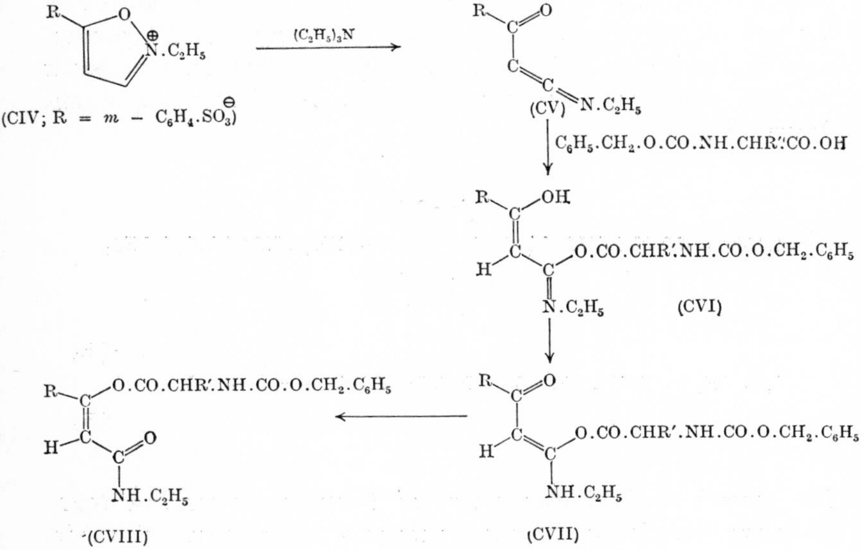

(CIV; R = $m-C_6H_4.SO_3^{\ominus}$) $\xrightarrow{(C_2H_5)_3N}$ R.C(=O).CH=C=N.C_2H_5 (CV)

(CV) $\xrightarrow{C_6H_5.CH_2.O.CO.NH.CHR'.CO.OH}$ R.C(OH)=CH.C(O.CO.CHR'.NH.CO.O.CH_2.C_6H_5)=N.C_2H_5 (CVI)

(CVI) $\rightarrow$ R.C(=O).CH=C(O.CO.CHR'.NH.CO.O.CH_2.C_6H_5).NH.C_2H_5 (CVII)

(CVII) $\rightarrow$ R.C(O.CO.CHR'.NH.CO.O.CH_2.C_6H_5)=CH.C(=O).NH.C_2H_5 (CVIII)

The isoxazolium salt (CIV), in presence of tertiary base, undergoes ring-opening to give an α-ketoketenimine (CV). The latter reacts with an *N*-acylamino acid by 1,4-addition to give initially the enol (CVI), which tautomerizes (CVI → CVII) and rearranges to the enolic ester (CVIII). This may be regarded as an activated ester and it reacts with esters of amino acids or peptides to give protected peptide derivatives.

$$\text{(CVIII)} \xrightarrow{NH_2.CHR''.CO.O.C_2H_5} C_6H_5.CH_2.O.CO.NH.CHR'.CO.NH.CHR''.CO.O.C_2H_5$$

$$+ \quad R{-}C({=}O){-}CH_2{-}C({=}O){-}NH.C_2H_5$$

$$(\text{CIX}; R = m - C_6H_4.SO_3^{\ominus})$$

The by-product (CIX), since it contains a sulphonic group, is soluble in water and therefore easily removed.

Two other types of active ester have recently been introduced. Esters of *N*-hydroxysuccinimide (CX), which can be synthesized from *N*-hydroxysuccinimide using a carbodi-imide (see p. 91), presumably owe their reactivity to the withdrawal of electrons by the two carbonyl groups. The solubility in water of *N*-hydroxysuccinimide, which is liberated during peptide synthesis, assists in isolation of the protected peptide (cf. the use of isoxazolium salts described above).

$$\underset{\text{(CX)}}{C_6H_5.CH_2.O.CO.NH.CHR.CO.O.N\begin{matrix} C(=O){-}CH_2 \\ | \\ C(=O){-}CH_2 \end{matrix}} + NH_2.CHR'.CO.O.C(CH_3)_3$$

$$\downarrow$$

$$C_6H_5.CH_2.O.CO.NH.CHR.CO.NH.CHR'.CO.O.C(CH_3)_3 + HO.N\begin{matrix} C(=O){-}CH_2 \\ | \\ C(=O){-}CH_2 \end{matrix}$$

The second method uses esters of *N*-hydroxypiperidine (CXI). Although these formally resemble esters of *N*-hydroxysuccinimide, their reactivity stems from a different cause. First, their reaction with amines is subject to acid catalysis. This may be explained if the heterocyclic nitrogen atom is protonated. Electron-withdrawal by $>\overset{\oplus}{N}H_2$ would increase the possibility of nucleophilic attack at the carbonyl-carbon atom.

$C_6H_5.CH_2.O.CO.NH.CHR.CO.O.N$(piperidine) (CXI) $\xrightarrow{H^{\oplus}}$ $C_6H_5.CH_2.O.CO.NH.CHR.CO.O.\overset{\oplus}{N}H$(piperidine)

$NH_2.CHR'.CO.O.CH_2.C_6H_5$

$C_6H_5.CH_2.O.CO.NH.CHR.CO.NH.CHR'.CO.O.CH_2.C_6H_5$ + HO.N(piperidine) + $H^{\oplus}$

A more subtle explanation may be necessary, however, since esters of *N*-hydroxypiperidine react with nucleophiles even in the absence of an added acid catalyst. It has been suggested that the heterocyclic nitrogen atom may be able to accept a proton from the incoming amine (CXII ⇌ CXIII).

R.CO.O.N(piperidine) + R'.NH_2 ⇌ R.C($O^{\ominus}$)—O—N(piperidine)···H—$\overset{\oplus}{N}$(H).R' (CXII) ⇌ R.C($O^{\ominus}$)—O—$\overset{\oplus}{N}$H(piperidine), R'.N(H) (CXIII) → R.C(=O).NHR' + $^{\ominus}O-\overset{\oplus}{N}H$(piperidine) ⟷ HO—N(piperidine)

Carbodi-imides, R.N:C:N.R', and in particular *NN'*-dicyclohexylcarbodi-imide (CXIV), have proved to be useful reagents for the synthesis of peptides both directly and indirectly through an intermediate active ester (Kurzer & Douraghi-Zadeh, 1967). Carboxylic acids react additively with carbodi-imides to give *O*-acylisoureas (CXV) which, in turn, react readily with nucleophiles such as amines and phenols to give amides and aryl esters respectively.

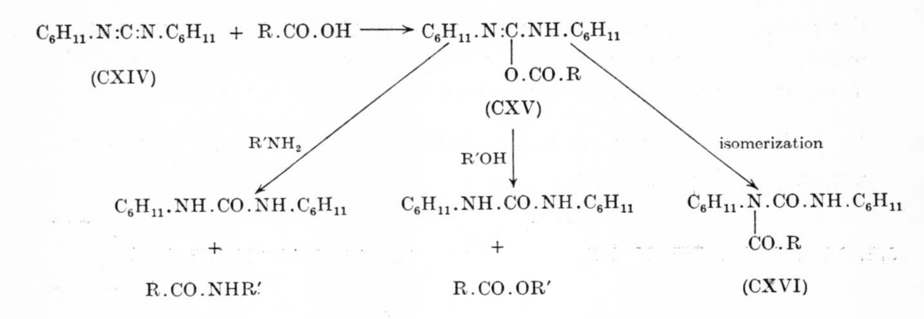

Isomerization of the *O*-acylisourea to the *N*-acylurea (CXVI) is an important side-reaction which lowers yields, since the *N*-acylurea is not susceptible to nucleophilic attack under mild conditions. Solvent and temperature influence the extent of the unwanted isomerization. Temperature should be kept at or below room temperature and the preferred solvents are tetrahydrofuran and methylene chloride. Clearly, the nature of the acid used to form the *O*-acylisourea and the nature of the nucleophile used to attack it will also help to determine if the reaction proceeds along the desired path. The unwanted product, *NN'*-dicyclohexylurea, is almost insoluble in tetrahydrofuran and methylene chloride and is therefore easily removed. Sometimes the insolubility of the starting materials requires that other solvents such as chloroform or *NN*-dimethylformamide be used. This always increases the risk that side-reactions and racemization will occur.

4.6. Racemization during peptide synthesis. Racemization is always a possible complication during the formation of a peptide bond and during any procedure which requires the use of alkaline conditions. Many of the properties, and especially the biological activity of peptides, depend on the optical configuration at the asymmetric carbon atoms. Some amino acids, such as threonine and isoleucine, have two asymmetric centres, but only one, the carbon atom adjacent to the carbonyl group, is usually susceptible to racemization during peptide synthesis. Racemization under alkaline conditions probably results from the reversible release of a proton from the α-carbon atom:

$$\mathrm{R.NH.CHR'.CO.R''} \rightleftharpoons \mathrm{R.NH.\overset{\ominus}{C}R'.CO.R''} + \mathrm{H^{\oplus}}$$

A number of factors including the nature of the α-*N*-acyl protecting group, the method of forming the peptide linkage, the presence of base, temperature, and solvent determine the possibility and extent of racemization during the formation of a peptide bond. The number of methods of forming peptide bonds and the number of different *N*-acyl protecting groups available reflect the amount of work and ingenuity which has been devoted to the problem of eliminating the risk of racemization. Many of the derivatives of *N*-acylamino acids, which are used in the synthesis of peptides, racemize by forming an intermediate oxazol-5-one (CXVII → CXVIII).

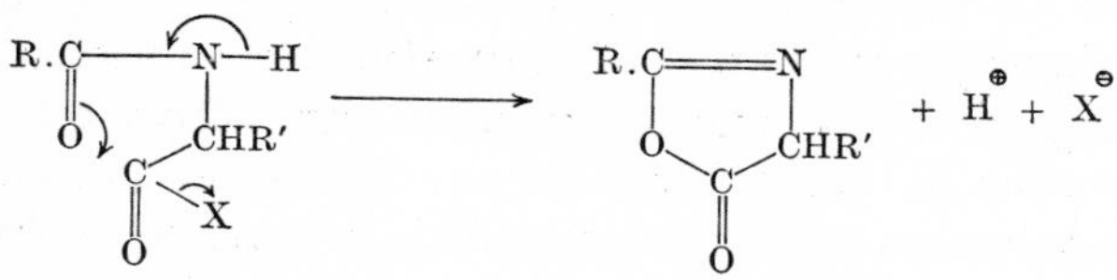

R.C—N—H, O, CHR′, C, X, O

(CXVII)

→

R.C═N, O, CHR′, C, O + $H^{\oplus}$ + $X^{\ominus}$

(CXVIII)

⇅

R.C═N, O, CR′, C, OH

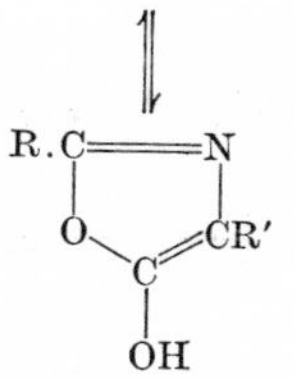

(CXIX)

The ability of the oxazol-5-one to form the tautomeric enol (CXIX) explains its facile racemization. It will be appreciated from a consideration of the above mechanism that formation of an oxazol-5-one with consequent racemization will be favoured by presence of base, use of polar solvents, pronounced carbonium character of the —CO.X group, good leaving characteristics of —X, and high electron density on the carbonyl-oxygen atom of the *N*-acyl group. It is obvious that the —CO.X group must have appreciable carbonium character and —X must be a good leaving group if the reagent is to be susceptible to nucleophilic attack by an ester of an amino acid or peptide. In choosing a reagent for synthesizing peptide bonds, it is necessary, therefore, to strike a balance between reactivity towards the nucleophilic amino group and the ease of formation of oxazol-5-one. It is not surprising then that the use of acid chlorides, for example, leads to extensive racemization. Protecting groups such as *N*-acetyl or *N*-benzoyl groups, apart from the difficulty of removing them at the end of a peptide synthesis, have a high electron density on the carbonyl-oxygen atom. Their use leads, therefore, to extensive racemization. In the urethane structure (e.g. *N*-benzyloxycarbonyl), on the other hand, the electron density appears to be highest on the nitrogen atom and so formation of an oxazol-5-one is not favoured. It is fortunate that the protecting groups which lead to the minimum degree of racemization are the easiest to remove.

Because the carbonyl-oxygen atom in a peptide bond carries an appreciable negative charge, the formation and use of *O*-acylisoureas, unsymmetrical acid anhydrides and active esters from *N*-acylpeptides is attended by a greater risk of racemization than when *N*-acylamino acids are used. This is true, of course, even when *N*-benzyloxycarbonylpeptides are used. For this reason du Vigneaud and his colleagues developed a stepwise method of peptide synthesis by adding one amino acid residue at a time at the *N*-terminus using the *p*-nitrophenyl ester of an *N*-acylated amino acid.

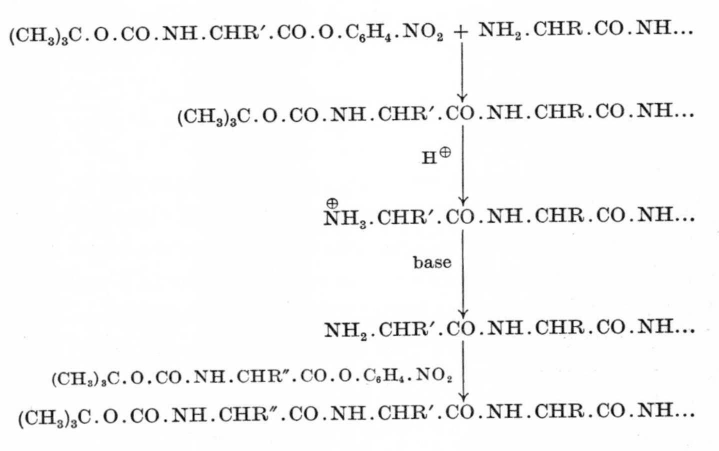

As mentioned above, the formation of a peptide bond through an acyl azide is free from the risk of racemization. This method has therefore retained its importance in spite of the inconvenience of having to make the hydrazide and of the possible side-reactions encountered when acyl azides are allowed to react with amines. Much less experience is available with the method of peptide synthesis which uses esters of *N*-hydroxypiperidine, but it also seems to give stereochemically pure products.

It is sometimes possible to plan a synthetic route so that fragments of the final peptide are synthesized by the stepwise procedure described above starting from a glycine ester or a proline ester (which resists racemization). The fragments will then have *C*-terminal glycine or proline and since racemization usually involves the *C*-terminal residue, the fragments can be coupled safely by any of the standard procedures of peptide bond synthesis.

It is useful to have sensitive methods for detecting if racemization has occurred during the synthesis of a peptide. A convenient practical assessment of the danger of racemization associated with a particular method for forming peptide bonds involves synthesizing *N*-benzyloxycarbonylglycyl-L-phenylalanylglycine ethyl ester from *N*-benzyloxycarbonylglycyl-L-phenylalanine or some suitable derivative and glycine ethyl ester. The racemic form of the protected tripeptide is readily separated from the L-enantiomer by fractional crystallization from ethanol. Less than 1 per cent of racemization can be detected in this way.

The stereospecificity of enzymes can be made use of to test for racemization during peptide synthesis. Portions of the synthetic peptide are digested with aminopeptidase (see p. 31) or one of the carboxypeptidases (see p. 32) and the hydrolysate is analysed at intervals for the presence of free amino acids by some suitable chromatographic procedure. From the known specificities of these enzymes, it is possible to predict the expected extent of degradation. If hydrolysis stops at an earlier stage than expected or if yields of amino acids suddenly drop, racemization is indicated. Alternatively, the synthetic peptide can be hydrolysed to free amino acids and the hydrolysate can be incubated with an amino acid oxidase (see p. 25) which cataylses the reaction:

$$\overset{\oplus}{N}H_3.CHR.CO.O^{\ominus} + O_2 + H_2O \rightarrow R.CO.CO.O^{\ominus} + NH_4^{\oplus} + H_2O_2$$

L-Amino acid oxidases occur in most snake venoms and many of these are commercially available. D-Amino acid oxidases occur in mammalian kidney. Although the enzymes are highly stereospecific, they display a much lower specificity towards the side-chain of the amino acids. The amino acid oxidase(s) present in kidney or snake venoms will therefore oxidize all the amino acids of the relevant configuration. If any racemization has occurred during the synthesis of a peptide, its acid hydrolysate will contain amino acids of the wrong configuration for one or other of the amino acid oxidases. After oxidation of the acid hydrolysate with amino acid oxidase, residual amino acids can be detected and determined by appropriate chromatographic procedures.

Finally, diastereoisomeric peptide derivatives can frequently be separated by gas-liquid chromatography. Peptides are converted into volatile derivatives such as *N*-trifluoroacetyl peptide esters. This is a sensitive procedure but it cannot be applied to large peptides because the volatility of their derivatives is too low.

4.7. Synthesis of α-melanocyte-stimulating hormone. To illustrate some of the principles described so far in this chapter, the synthesis of a naturally occurring peptide will be outlined. α-Melanocyte-stimulating hormone, which is found in the pituitary, stimulates pigmentation of the skin by melanin. The hormone is a tridecapeptide and has the sequence: *N*-Acetyl-Ser-Tyr-Ser-Met-Glu-His-Phe-Arg-Trp-Gly-Lys-Pro-Val-amide. The synthesis involved making three fragments, the *N*-terminal tetrapeptide (CXX), a hexapeptide containing residues 5–10 (CXXI) and the *C*-terminal tripeptide (CXXII), all with suitable protecting groups:

N-CH_3.CO-Ser-Tyr-Ser-Met-NH.NH_2
(CXX)

H-Glu-His-Phe-Arg-Trp-Gly-OH
|
O.$C(CH_3)_3$ (on Glu)
(CXXI)

O.CO.$C(CH_3)_3$ (on Lys)
|
C_6H_5.CH_2O.CO-Lys-Pro-Val-NH_2
(CXXII)

The peptide derivative (CXX) was converted into the azide and coupled with the peptide (CXXI) to give the decapeptide derivative (CXXIII):

N-CH_3.CO-Ser-Tyr-Ser-Met-Glu-His-Phe-Arg-Trp-Gly-OH
|
O.$C(CH_3)_3$ (on Glu)
(CXXIII)

The α-amino group of the peptide derivative (CXXII) was selectively uncovered by hydrogenolysis. (Notice the use of different protecting groups on the amino groups of lysine.) The decapeptide derivative (CXXIII) and the tripeptide containing a free α-amino group were now coupled by using *NN'*-dicyclohexylcarbodi-imide. (Notice that this method was used when glycine was *C*-terminal but that the azide route was used in the synthesis of the decapeptide when methionine was *C*-terminal.) The t-butyl ester and *N*-t-butyloxycarbonyl protecting groups were removed with trifluoroacetic acid to give α-melanocyte-stimulating hormone. The stereochemical purity was demonstrated by acid hydrolysis followed by digestion with L-amino acid oxidase (Schwyzer, Costopanagiotis & Sieber, 1963).

4.8. Cyclic peptides. Cyclic peptides may be classified into homodetic and heterodetic types. The former may be considered to be derived from a corresponding linear peptide by intramolecular formation of a peptide bond between the *N*-terminal amino group and the *C*-terminal carboxyl group. In heterodetic peptides, on the other hand, cyclization is effected through other types of linkage such as ester or disulphide bonds. Apart from any academic interest in the synthesis of cyclic peptides, many naturally-occurring peptides are either homodetic or heterodetic cyclic peptides.

The synthesis of a homodetic cyclic peptide is similar in principle to the synthesis of a linear peptide. It is usual to make a linear peptide (CXXIV) containing an active ester group, and a masking group for the α-amino group which can be removed under very mild conditions. The protecting group at the *N*-terminus is removed under acidic conditions so that the liberated amino group is protonated (CXXV). Under these conditions, the peptide ester is prevented from undergoing polymerization or cyclization. In very dilute solution (to diminish intermolecular reaction), the ester (CXXV) is treated with a base such as pyridine to remove the proton and allow intramolecular acylation to proceed to give the cyclic peptide (CXXVI).

There is always competition between intramolecular and intermolecular reaction and yields of cyclic product are often low. Cyclization of derivatives of dipeptides to 1,4-dioxopiperazines

(CXXVII), however, occurs extremely readily and even alkyl esters of dipeptides undergo cyclization. The reason for this facile cyclization lies in the formation of a six-membered ring, which is stable even though the peptide bonds have to assume a *cis*-conformation (p. 50).

$$(CH_3)_3C.O.CO.NH.CHR_1.CO.NH.CHR_2.CO\ldots\ldots NH.CHR_n.CO.O.C_6H_4.NO_2$$

(CXXIV)

$\downarrow H^{\oplus}$

$$\overset{\oplus}{N}H_3.CHR_1.CO.NH.CHR_2.CO\ldots\ldots NH.CHR_n.CO.O.C_6H_4NO_2$$

(CXXV)

$\downarrow$ base

$$\underline{NH.CHR_1.CO.NH.CHR_2.CO\ldots\ldots NH.CHR_n.CO}$$

(CXXVI)

N H, R2, H, O, H, R1, O, N H

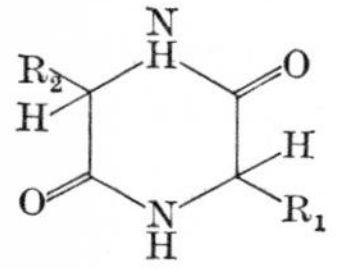

(CXXVII)

Attempted synthesis of a cyclotripeptide usually leads to the formation of a cyclohexapeptide by the so-called 'doubling reaction'. The linear sequence of amino acids occurs twice in the cyclic peptide. Thus, cyclization of the tripeptide ester (CXXVIII) and the hexapeptide ester (CXXIX) give the same cyclohexapeptide (CXXX). (Note that, when using abbreviations for amino acid residues in cyclic peptides, it is conventional to indicate the direction of the peptide bond, CO → NH, by an arrow.)

Only a small number of cyclotetrapeptides and cyclopentapeptides have been synthesized. In many attempted syntheses, the 'doubling reaction' occurred. This phenomenon seems to be energetically favoured by the increased possibilities of preserving *trans* peptide linkages and of forming transannular hydrogen bonds. The side-chains of the amino acids also seem to exert a steric effect. The last effect can lead to stereospecific cyclization; for instance, the racemic tripeptide ester (CXXXI) gives, as the major product of cyclization, the *meso*-cyclohexapeptide (CXXXII).

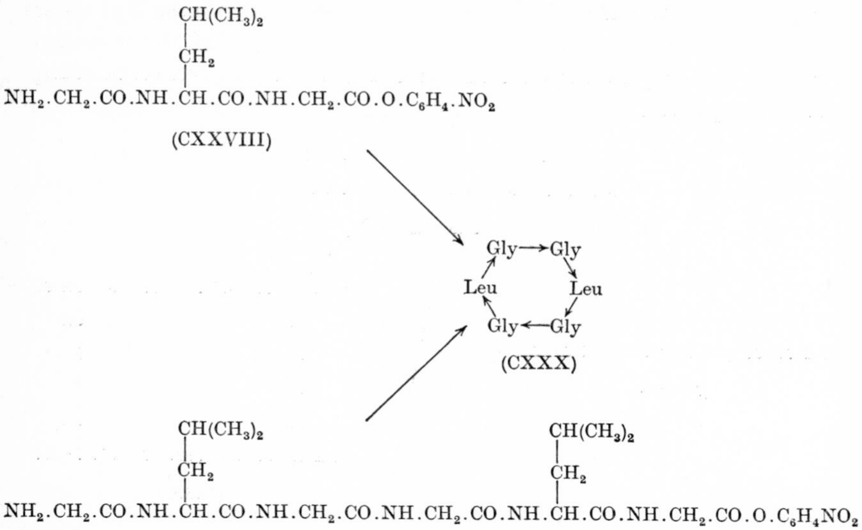

$$\begin{array}{l} \qquad\qquad\quad CH(CH_3)_2 \\ \qquad\qquad\quad | \\ \qquad\qquad\quad CH_2 \\ \qquad\qquad\quad | \\ NH_2.CH_2.CO.NH.CH.CO.NH.CH_2.CO.O.C_6H_4.NO_2 \end{array}$$

(CXXVIII)

Gly→Gly
↗ ↘
Leu Leu
↖ ↙
Gly←Gly

(CXXX)

$$\begin{array}{l} \qquad\qquad\quad CH(CH_3)_2 \qquad\qquad\qquad\qquad\qquad\qquad\qquad\quad CH(CH_3)_2 \\ \qquad\qquad\quad | \qquad\qquad\qquad\qquad\qquad\qquad\qquad\qquad\qquad\quad | \\ \qquad\qquad\quad CH_2 \qquad\qquad\qquad\qquad\qquad\qquad\qquad\qquad\qquad CH_2 \\ \qquad\qquad\quad | \qquad\qquad\qquad\qquad\qquad\qquad\qquad\qquad\qquad\quad | \\ NH_2.CH_2.CO.NH.CH.CO.NH.CH_2.CO.NH.CH_2.CO.NH.CH.CO.NH.CH_2.CO.O.C_6H_4NO_2 \end{array}$$

(CXXIX)

$$\begin{array}{l} \qquad\qquad\quad CH_2.C_6H_5 \\ \qquad\qquad\quad | \\ NH_2.CH_2.CO.NH.CH.CO.NH.CH_2.CO.O.C_6H_4.NO_2 \end{array}$$

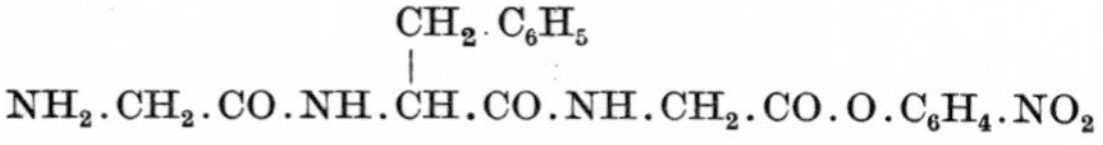

(CXXXI)

Gly→Gly
↗ ↘
D-Phe L-Phe
↖ ↙
Gly←Gly

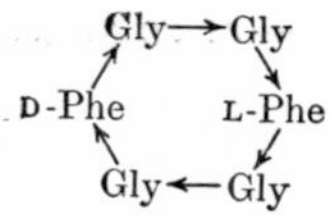

(CXXXII)

Heterodetic cyclic peptides containing a disulphide linkage are usually synthesized by oxidation of an open-chain peptide containing two cysteinyl residues (CXXXIII). Several cyclic products (CXXXIV–CXXX) as well as linear polymers are theoretically possible, and the distance between the cysteinyl residues determines which predominates.

The available evidence seems to indicate that when $n \leqslant 3$, the antiparallel cyclopeptide (CXXXVI) is the main product apart from linear polymers. On the other hand, when $n = 4-6$, the cyclopeptide (CXXXIV) is preferentially formed. Disulphide loops of the type (CXXXIV; $n = 4$) are present in oxytocin, vasopressin and insulin, and the chemical synthesis of these hormones and their analogues is simplified by the ready formation of the twenty-membered ring by intramolecular oxidation of the corresponding linear peptides.

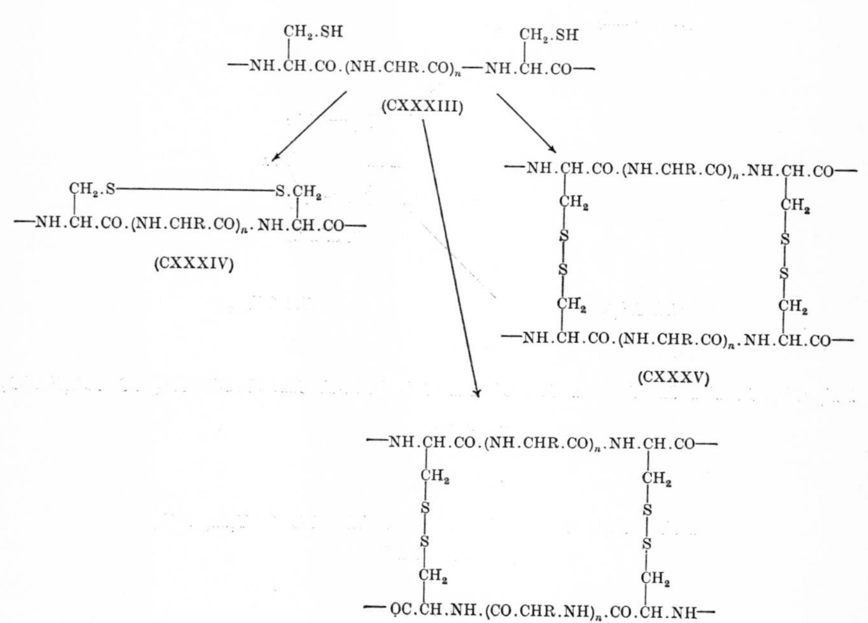

Heterodetic cyclic peptides containing an ester linkage (lactones) can arise by esterification of the hydroxyl group of serine or threonine with the carboxyl group of another amino acid residue. Such structures occur quite widely in peptide antibiotics, but few have been synthesized (Russell, 1966).

4.9. Synthesis of peptides on a solid support. Recently, Merrifield (1965) has developed a synthetic procedure in which a peptide is built up by one amino acid at a time using an insoluble synthetic resin as a support for the peptide. The peptide is covalently linked to the resin and is liberated at the end of the synthesis. The resin is a copolymer of styrene and divinylbenzene in which some of the aromatic rings are chloromethylated and nitrated (CXXXVII). A suitable *N*-acyl derivative (e.g. t-butyloxy-carbonyl) derivative of the *C*-terminal amino acid of the peptide to be synthesized is heated under reflux with the resin and tertiary base in a suitable solvent to convert it into an insoluble benzyl ester derivative (CXXXVIII). The *N*-acyl group is removed (CXXXIX) and a dipeptide derivative (CXL) is formed by coupling an *N*-acyl-amino acid using *NN′*-dicyclohexylcarbodi-imide. The polypeptide chain is extended by a similar cycle of reactions. At the end of the synthesis, the *N*-acyl protecting group can be removed and the ester bond linking the peptide to the resin can be cleaved with hydrogen bromide in trifluoroacetic acid.

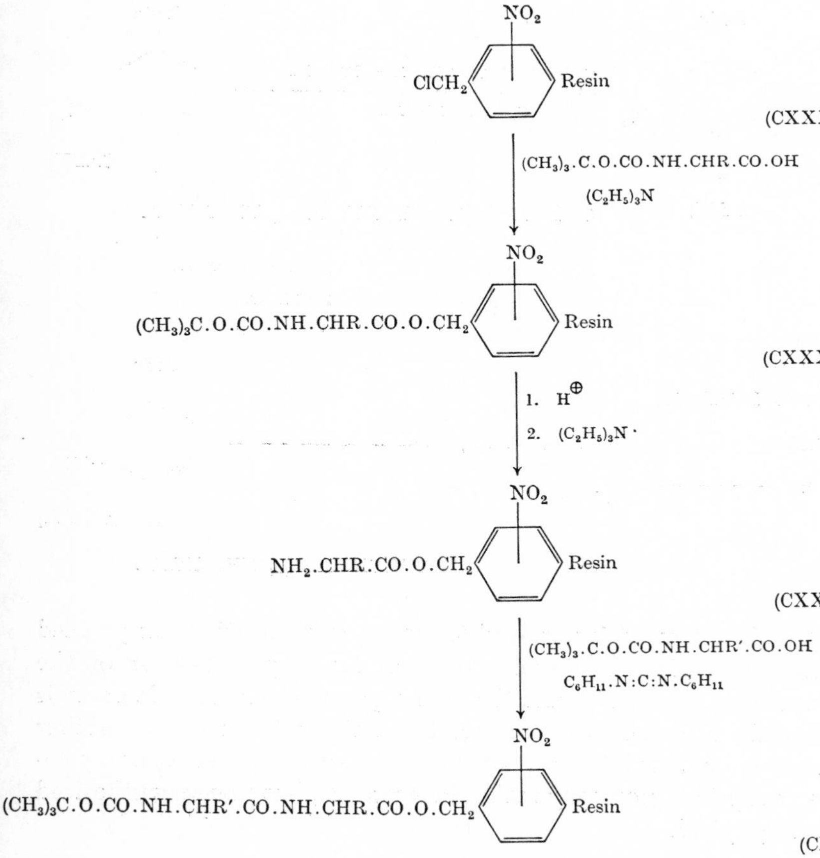

NO_2

$ClCH_2$ — Resin

(CXXXVII)

↓ $(CH_3)_3.C.O.CO.NH.CHR.CO.OH$

$(C_2H_5)_3N$

NO_2

$(CH_3)_3C.O.CO.NH.CHR.CO.O.CH_2$ — Resin

(CXXXVIII)

↓ 1. $H^{\oplus}$

2. $(C_2H_5)_3N$

NO_2

$NH_2.CHR.CO.O.CH_2$ — Resin

(CXXXIX)

↓ $(CH_3)_3.C.O.CO.NH.CHR'.CO.OH$

$C_6H_{11}.N:C:N.C_6H_{11}$

NO_2

$(CH_3)_3C.O.CO.NH.CHR'.CO.NH.CHR.CO.O.CH_2$ — Resin

(CXL)

By using a solid phase to support the peptide and intermediate products, the need to purify the latter by crystallization is avoided, since unused reagents can be washed away between steps. Syntheses can thus be completed quickly and in high yield. For example, Merrifield obtained the nonapeptide, bradykinin, in 68 per cent overall yield in eight days. More recently, Merrifield has mechanized the procedure so that the requisite reagents can be added automatically at predetermined times.

A related procedure has been used for the synthesis of cyclopeptides. A cross-linked poly-4-hydroxy-3-nitrostyrene (CXLI) and an *N*-acylpeptide give an insoluble aryl ester (CXLII) in presence of *NN'*-dicyclohexylcarbodi-imide.

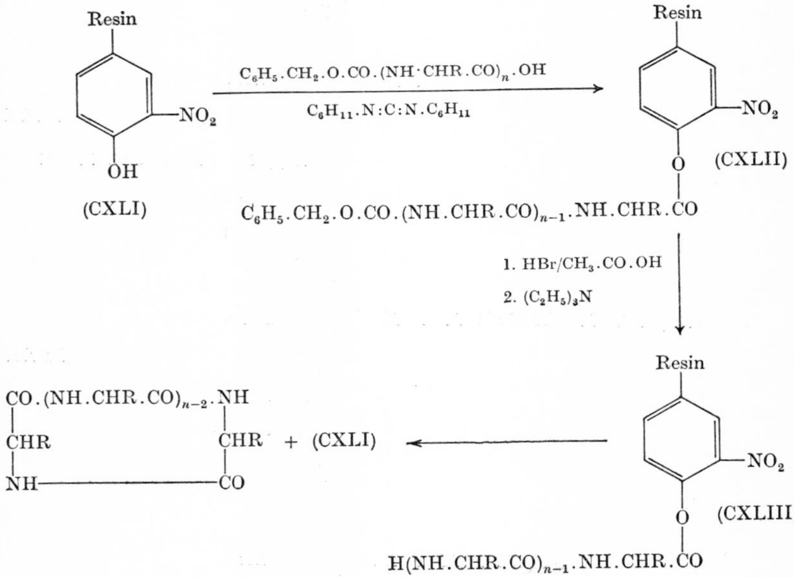

Resin

$-NO_2$

OH

(CXLI)

$\xrightarrow[C_6H_{11}.N:C:N.C_6H_{11}]{C_6H_5.CH_2.O.CO.(NH\cdot CHR.CO)_n.OH}$

Resin

$-NO_2$

O (CXLII)

$C_6H_5.CH_2.O.CO.(NH.CHR.CO)_{n-1}.NH.CHR.CO$

1. $HBr/CH_3.CO.OH$

2. $(C_2H_5)_3N$

$\downarrow$

Resin

$-NO_2$

O (CXLIII)

$H(NH.CHR.CO)_{n-1}.NH.CHR.CO$

$\longleftarrow$

$CO.(NH.CHR.CO)_{n-2}.NH$

CHR CHR + (CXLI)

NH——CO

The *N*-acyl group is removed (CXLIII) and cyclization can proceed with little risk of the formation of linear polymers or of the 'doubling reaction', since the amino group of one peptide chain is unlikely to be able to approach the active ester group of another peptide chain while they are attached to the resin. Some cyclotetrapeptides, normally rather inaccessible, have been synthesized in high yield.

5. Biosynthesis of proteins

5.1. Prebiological origin of proteins. It is not possible to consider in detail here the various theories which have been advanced to account for the appearance of life in the universe. We shall confine our attention to examining likely mechanisms of synthesis of amino acids and proteins which could have operated on earth and made possible the appearance of life. Thirty years ago, Oparin suggested that the formation of the first living organisms involved three main stages: (*a*) synthesis of simple compounds such as amino acids at a time when the earth had a reducing atmosphere containing hydrogen, methane, ammonia, nitrogen, water vapour, hydrogen sulphide and later hydrogen cyanide; (*b*) formation of a hot primaeval 'soup' containing simple organic and inorganic compounds in which macromolecules such as proteins were formed; (*c*) emergence of discrete multimolecular systems which later gave rise to cellular organisms.

A simple experiment by Miller (1955) revealed that stage (*a*) was feasible in a primordial atmosphere provided sufficient energy was available as it almost certainly would have been. A mixture of methane, ammonia, hydrogen and water vapour was circulated in an apparatus past an electrical discharge from tungsten electrodes. After a few days, the condensate was found to contain glycine, alanine, aspartic acid and glutamic acid together with a number of other carboxylic acids. It was also shown that hydrogen cyanide was formed at an early stage and reached a steady concentration, but gradually disappeared later on. Other experiments showed that hydrogen could be replaced by carbon monoxide. Heat could replace the electric discharge as a source of energy; when a mixture of methane, ammonia and water was passed through a silica tube at 900–1000 °C, fourteen α-amino acids were formed. It is important to understand that stage (*a*) required a reducing atmosphere, but that life as we know it could not emerge until the atmosphere contained oxygen and was essentially free of gases such as hydrogen cyanide, ammonia and carbon monoxide. The presence of hydrogen cyanide in the primitive atmosphere is important, however, and it is possible that amino acids arose from a Strecker reaction involving aldehydes, ammonia and hydrogen cyanide. Alternatively, reaction of ammonium cyanide in hot aqueous solution gives many amino acids together with some polypeptides.

A number of other theories to account for the formation of polypeptides from amino acids have been advanced and supported by model experiments. For example, when ammonium salts of amino acids are heated under anhydrous conditions, or when an excess of glutamic acid is heated with other amino acids, polymeric products are formed. In hot aqueous solution, amides of amino acids polymerize. A recent attractive theory invokes dicyandiamide as the reagent for producing polypeptides from amino acids. Dicyandiamide, which was shown to be more effective than cyanamide in model experiments, exists to some extent as the tautomeric diimide, $NH_2.C(:NH).N:C:NH$, and recalls the use of carbodiimides in the synthesis of peptides. A different approach envisages the formation of polyglycine or similar precursor and the subsequent introduction of side-chains corresponding to the more complex amino acids.

Attempts have been made to explain the emergence of optically active polypeptides. It has been suggested, for example, that the circularly polarized light reflected from the moon or the plane polarized light reflected from the sea would favour asymmetric synthesis. A second hypothesis suggests that polypeptide formation might have been catalysed at the surface of clays and it is possible that one stereochemical form would be found preferentially. Finally, attempts have been made to explain the appearance of optically asymmetric molecules by invoking the theory of parity non-conservation (Ulbricht, 1959). In one recent hypothesis (Yamagata, 1966) it is suggested that as a result of the non-conservation of parity, D- and L-forms of a molecule have slightly different wave-functions and hence they have slightly different bond energies and probabilities of chemical reaction. Although the difference might be minute for a single asymmetric centre, it is suggested that for large molecules with many asymmetric centres, the probability that only one form would emerge from a synthesis could be quite high.

The postulated third stage for the genesis of life according to Oparin's theory involves the formation of multimolecular systems. Molecules of proteins have a pronounced tendency to undergo aggregation due to the possibility of forming noncovalent bonds. Oparin has suggested that polypeptides and other macromolecules may have formed complex coacervates. For example, protein molecules of opposite charge could be attracted to one another enclosing a sheath of water between them. Such complex coacervates might be expected to group in an organized fashion and assume some of the functions normally associated with the cell. Model experiments have shown that coacervates can selectively absorb and concentrate other molecules from the external medium. If the coacervate contains an enzyme, its substrate and the products may accumulate in the coacervate. For example, if a coacervate of gum arabic and histone also containing phosphorylase is placed in a solution containing glucose-1-phosphate, starch accumulates in the coacervate.

It is clear from this brief account that there is no dearth of plausible hypotheses, usually supported by model experiments, which attempt to explain the prebiological formation of proteins as part of organized systems from which living organisms could have evolved. Not surprisingly, there is a disconcerting lack of concrete evidence and it is possible that space exploration will reveal the existence of prebiological states or primitive forms of extraterritorial life which will throw more light on the way in which life evolved on earth. Already, careful analysis has revealed evidence, albeit disputed, of organic matter in an organized state in meteorites.

For further consideration of this fascinating field, the reader is referred to reviews and books by Keosian (1964), Oparin (1965), and Pattee (1965).

5.2. The biosynthesis of proteins. The belief that nucleic acids are closely concerned with protein biosynthesis has long been held. In particular, it was thought probable that the genetic information required to produce proteins with the correct amino acid sequence resided in molecules of deoxyribonucleic acid (DNA) located in chromosomes. In addition, it was known that one molecule of adenosine 5′-triphosphate is converted into adenosine

5′-phosphate for each peptide bond which is synthesized. Only during the last ten years, however, has the mechanism of protein biosynthesis been unravelled. The amount of work which has been required and the rate at which facts and theories have been published make it impossible to do more than give a bare outline of the mechanism.

Genetic information concerning the sequence of amino acids in a protein is normally stored in one strand of a doubly-stranded DNA molecule in the chromosomes found in the cell nucleus. The primary structure of a protein to be synthesized is determined by the base sequence of the DNA molecule. The actual synthesis of the protein usually occurs outside the cell nucleus on small particles known as ribosomes. The information concerning the sequence of amino acids is transferred from the cell nucleus to the ribosomes by a single-stranded messenger ribonucleic acid (mRNA). The DNA molecule acts as a template for the synthesis of mRNA and the correct sequence is determined by base-pairing. Watson and Crick showed that, in double-stranded DNA, adenine in one chain is hydrogen-bonded to thymine in the other chain; likewise, guanine is paired with cytosine. A similar situation is believed to exist when a single strand of DNA acts as a template for the synthesis of mRNA with the exception that adenine in DNA pairs with uracil in RNA and thymine in DNA pairs with adenine in RNA.

We must now consider briefly how the sequence of bases in a nucleic acid determine the sequence of amino acids in a protein. This is referred to as the genetic code (Crick, 1963, 1966; Bennett & Dreyer, 1964), and in a brilliant series of experiments, Crick and his colleagues showed that three consecutive nucleotides code for one amino acid in a protein. Such triads of nucleotides in mRNA are referred to as codons. There are 4^3 ways of selecting a triad of nucleotides from the four bases, guanine (G), adenine (A), cytosine (C), and uracil (U), which occur in mRNA. Since, with a few exceptions such as hydroxyproline and hydroxylysine, there are only twenty different amino acids in proteins, it would seem likely that the code is degenerate, i.e. more than one triad is capable of coding for the same amino acid. The existence of degeneracy in the genetic code may be biologically advantageous, since not all mutations, associated with change of nucleotide sequence in DNA or errors in transcription in mRNA, will lead to a change of primary structure in the resultant protein. In chapter 6, it is shown that

TABLE 5.1. *The genetic code*

1st base of triad	2nd base of triad: U	C	A	G	3rd base of triad
U	Phe	Ser	Tyr	CySH	U
	Phe	Ser	Tyr	CySH	C
	Leu	Ser	Nonsense	Nonsense	A
	Leu*	Ser	Nonsense	Trp	G
C	Leu	Pro	His	Arg	U
	Leu	Pro	His	Arg	C
	Leu	Pro	Glu(NH_2)	Arg	A
	Leu	Pro	Glu(NH_2)	Arg	G
A	Ile	Thr	Asp(NH_2)	Ser	U
	Ile	Thr	Asp(NH_2)	Ser	C
	Ile	Thr	Lys	Arg	A
	Met*	Thr	Lys	Arg	G
G	Val	Ala	Asp	Gly	U
	Val	Ala	Asp	Gly	C
	Val	Ala	Glu	Gly	A
	Val	Ala	Glu	Gly	G

* These triads apparently code for *N*-terminal *N*-formylmethionine in *E. coli*.

degeneracy and redundancy can exist in proteins, so that the possibility exists, at two levels, of preserving the biological activity of polypeptides in spite of mutations. The genetic code is given in table 5.1, and it will be seen that much of the degeneracy concerns the third base of a triad. For example serine is coded by the triads UCU, UCC, UCA and UCG. It is noticeable that coding for those amino acids which occur most frequently in proteins shows considerable degeneracy. For example, the commonly occurring amino acids, glycine, alanine, valine, leucine, arginine, serine and threonine each have four to six different codons. Conversely, the rarer amino acids, tryptophan and methionine have only one or two codons. There is another feature worth noting about the genetic code; amino acids with similar side chains frequently have similar codons. Thus, even if a mutation does occur which is non-degenerate, it is quite possible that it will produce a protein with a similar charge distribution and conformation which might function adequately. For instance, if GAU (which codes for aspartic acid) were replaced by GAA, a similar amino acid, glutamic acid, would be produced. As a second example, the triads UUU, CUU,

AUU and GUU all code for amino acids with hydrophobic side chains (phe, leu, ile, val).

There are a few final points concerning the genetic code which should be mentioned. First, the genetic code is non-overlapping, i.e. each base can code in only one triad. Secondly, the triads AUG and UUG appear to code for *N*-terminal *N*-formyl-L-methionine in *E. coli* and thus act as chain initiators. The *N*-formyl- or *N*-formylmethionyl groups may subsequently be removed enzymatically. A 'nonsense codon' may terminate the synthesis of a protein. Codons to initiate and stop chain synthesis are necessary if, as seems likely, one mRNA molecule is able to code for more than one protein molecule. Thirdly, the internucleotide linkage in ribonucleic acids is known to be a phosphodiester involving the hydroxyl group at $C_{3'}$ on one nucleotide and the hydroxyl group at $C_{5'}$ on the next. Because this linkage is unsymmetrical, there are two possible ways in which the codons can be 'read'. Thus, a segment of mRNA, which contains the AAC codon for asparagine, could have one of two structures (CXLIV and CXLV). In fact, experiment has shown that the codons are read as in the former structure.

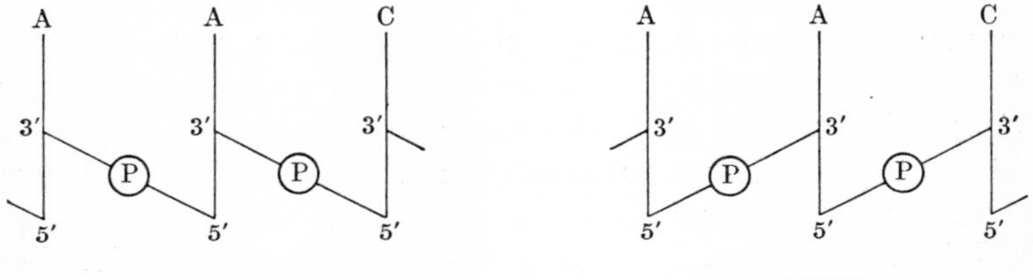

A A C
3′ 3′ 3′
P P
5′ 5′ 5′

(CXLIV)

A A C
3′ 3′ 3′
P P
5′ 5′ 5′

(CXLV)

We must now consider events during protein synthesis on the ribosomes. The latter become attached to mRNA at specific points on the nucleic acid molecules. These are the nucleotide sequences which code for the *N*-terminal amino acid, since it is known that synthesis proceeds by the stepwise addition of amino acids to the *C*-terminus of a growing polypeptide chain. With the addition of each amino acid residue, the mRNA and ribosome move relative to one another until the protein chain is complete. Only a small section of the mRNA molecule is in contact with a ribosome at any one time. It is possible, therefore, for several ribosomes to be attached to a single mRNA molecule at once so that several identical polypeptide chains are being synthesized simultaneously although, at any instant, they will be in different states of completion. A collection of ribosomes all functioning with the same mRNA molecule are known as a polyribosome or polysome.

The way in which mRNA directs that the correct amino acids are incorporated into the protein chain and the mechanism of formation of peptide bonds are related aspects of the problem which we can deal with together. The incorporation of each amino acid requires a specific transfer RNA (tRNA). These are small, soluble nucleic acids containing about eighty nucleotide units. As well as being much smaller than most nucleic acids, tRNA's contain unusual nucleotides such as pseudodouridylic acid, inosinic acid and a variety of methylated derivatives. The sequence of nucleotides of a few tRNA's from yeast is already known. All the tRNA molecules terminate with the sequence CCA and it is here that an amino acid is attached. There is probably a different tRNA for each codon. The conformations of tRNA's have not been elucidated, but it is believed that the molecules are looped into a structure containing a double helix with pairing of most of the nucleotides by hydrogen bonds. It is further believed that three bases in the tRNA (the anticodon), which are not hydrogen-bonded in tRNA itself, are bound to the codon on the mRNA. Specificity in the anticodon-codon linkage could be explained if hydrogen bonds, similar to those known to exist in double-stranded DNA, are formed between complementary purine-pyrimidine pairs of bases.

Amino acids, before they become attached to tRNA, react with adenosine triphosphate to give an aminoacyl adenylate (CXLVI) and pyrophosphate. Although phosphate groups are shown as anions, in practice, the negative charges are probably shielded by inorganic cations such as Mg^{2+}. The aminoacyl adenylate remains bound to the enzyme and the amino acid is then transferred to the tRNA molecule (CXLVII) in a second reaction.

Because the biosynthesis of a protein proceeds in a stepwise manner, it is reasonable to suppose that molecules of tRNA only become attached to mRNA when they carry an amino acid or peptide. It may be that the attachment of an amino acid of tRNA produces a change of conformation which uncovers the anticodon. Presumably, only two tRNA molecules are attached to the mRNA and ribosome at any one time; the amino acid, which will be the *N*-terminal residue, is transferred to the adjacent tRNA carrying

$$\text{Adenosine.O.}\overset{\overset{\displaystyle O}{\|}}{\underset{\underset{\displaystyle O^{\ominus}}{|}}{P}}\text{.O.}\overset{\overset{\displaystyle O}{\|}}{\underset{\underset{\displaystyle O^{\ominus}}{|}}{P}}\text{.O.}\overset{\overset{\displaystyle O}{\|}}{\underset{\underset{\displaystyle O^{\ominus}}{|}}{P}}\text{.O}^{\ominus} + \overset{\oplus}{N}H_3\text{.CHR.CO.O}^{\ominus} \xrightarrow{\text{Aminoacyl synthetase}} {}^{\ominus}O\text{—}\overset{\overset{\displaystyle O}{\|}}{\underset{\underset{\displaystyle O^{\ominus}}{|}}{P}}\text{.O.}\overset{\overset{\displaystyle O}{\|}}{\underset{\underset{\displaystyle O^{\ominus}}{|}}{P}}\text{.O}^{\ominus}$$

$$+ \text{Adenosine.O.}\overset{\overset{\displaystyle O}{\|}}{\underset{\underset{\displaystyle O^{\ominus}}{|}}{P}}\text{.O.CO.CHR.}\overset{\oplus}{N}H_3$$

(CXLVI)

$$\text{Adenosine.O.}\overset{\overset{\displaystyle O}{\|}}{\underset{\underset{\displaystyle O^{\ominus}}{|}}{P}}\text{.O.CO.CHR.}\overset{\oplus}{N}H_3 \quad + \quad \text{[terminal nucleotide of tRNA: } {}^{\ominus}O\text{–P(=O)–O-ribose(2'-OH, 3'-OH)–Adenine]}$$

(CXLVI) (terminal nucleotide of tRNA)

Adenine

OH OH

↓

$$\text{Adenosine.O.}\overset{\overset{\displaystyle O}{\|}}{\underset{\underset{\displaystyle O^{\ominus}}{|}}{P}}\text{.O}^{\ominus} \quad + \quad \text{[}{}^{\ominus}O\text{–P(=O)–O-ribose–Adenine, with } \overset{\oplus}{N}H_3\text{.CHR.CO.O on one ribose hydroxyl, OH]} \quad + \quad H^{\oplus}$$

Adenine

$\overset{\oplus}{N}H_3$.CHR.CO.O OH

(CXLVII)

the next amino acid in the sequence. When the peptide bond is formed, the first tRNA is detached from the mRNA and the second tRNA now bears the *N*-terminal dipeptide sequence. The next tRNA loaded with the third amino acid is bound to the mRNA and the dipeptide on the second tRNA is transferred to the third tRNA. The tRNA's corresponding to the amino acids near the *C*-terminus thus carry an ever growing protein chain until a codon is reached which signals the end of the chain. The completed protein molecule is then detached. This hypothesis is depicted diagrammatically in fig. 5.1. The formation of peptide bonds is enzymatically controlled and a molecule of guanosine triphosphate is required for the synthesis of each peptide bond, although its precise function in the reaction mechanism is unknown.

The above account of protein biosynthesis has been greatly

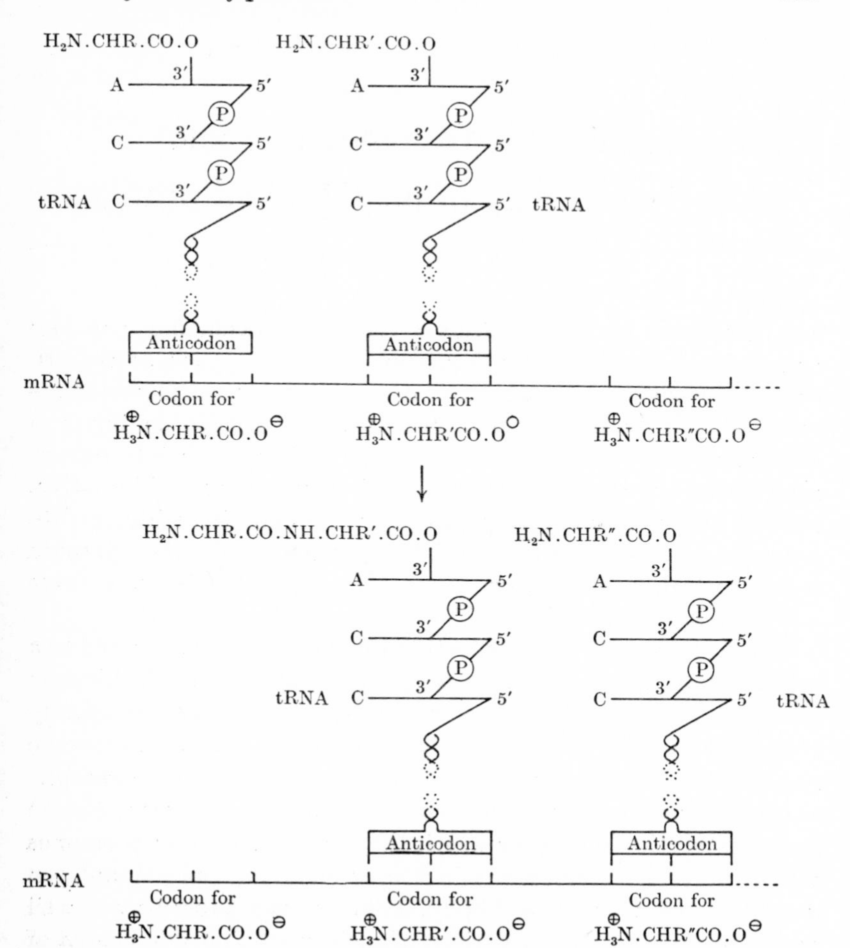

Fig. 5.1. Diagrammatic representation of a proposed mechanism of protein biosynthesis. The amino acid borne by the tRNA at the left-hand side of the upper part of the diagram is transferred to the second tRNA (lower part of diagram). The first tRNA is detached and the tRNA bearing the third amino acid is then attached to the mRNA in readiness for the synthesis of the second peptide bond.

simplified and even distorted in the sense that all genetical considerations have been ignored. Fuller accounts of the work leading to the elucidation of the genetic code and the elaboration of the current views of protein biosynthesis have been given by Crick (1963), Arnstein (1965), Stretton (1965) and Watson (1965).

6. Relationships between structure and biological activity of peptides and proteins

6.1. General considerations. Naturally occurring peptides and proteins range in size from small peptides, such as the tripeptide glutathione, to proteins with molecular weights of more than 10^6. Their biological activities cover almost as broad a spectrum. Some are relatively inert (e.g. keratin) while others, such as the polypeptide hormones and enzymes are highly active. Not all biologically active peptides are favourable to life; micro-organisms are sensitive to peptide antibiotics and the protein toxin from *Clostridium botulinum* is one of the most lethal substances known towards animals.

Manifestation of biological activity can ultimately be traced to a perturbation, which may be catalytic or inhibitory, of the kinetics of one or more biochemical reactions. Moreover, most biologically active compounds owe their activity to their ability to form some sort of complex or compound with cell constituents. For example, enzymes form a loose complex with a substrate as an essential step in the catalytic process for which they are responsible; hormones are bound to some receptor which may be an enzyme; an antigen is firmly bound by an antibody. Peptides and proteins are well suited to interact with a variety of molecules. The side chains of the constituent amino acids can participate in the formation of hydrogen bonds, hydrophobic bonds and salt linkages. Moreover, a protein molecule may be folded in such a manner as to provide topographical features which allow a precise fit with another molecule such as a substrate. The overall shape of the protein molecule may depend on the nature and sequence of most of the amino acids which it contains.

Enzymes, in addition to binding a substrate, catalyse such reactions as hydrolysis, synthesis and cleavage of carbon–carbon bonds, redox reactions and phosphorylations. Some enzymes contain a prosthetic group such as haem or require a coenzyme such as nicotinamide-adenine dinucleotide. Whether such non-peptide moieties are present or not, some of the amino acids in the peptide chain must play a part in the catalytic mechanism. This part of the molecule is often referred to as the *catalytic site*. Other amino acids, which presumably adjoin or overlap the catalytic site, form the *binding site*; this part of the molecule is responsible for binding the substrate(s) and determines the specificity of the enzyme. Collectively, the catalytic and binding sites are usually referred to as the *active centre* of the enzyme. The amino acids which constitute the active centre may not be adjacent to one another in the primary structure of the protein, but they must then be spatially juxtaposed by folding of the molecule. Many of the amino acids in an enzyme molecule must have a passive role and help to maintain the correct spatial relationship between the various parts of the active centre.

As will be shown later, many biologically active peptides and proteins can be chemically modified, either by substitution of a group or by removal of some amino acids, without altering the biological activity. Clearly, such molecules contain amino acids which are redundant. It may be wondered why living organisms synthesize molecules which are more complex than necessary for the function which they perform. One obvious possibility is that the larger molecules are less likely to diffuse away through cell walls. Again, it is possible that the added complexity provides some protection against degradation by the ubiquitous proteolytic enzymes. There is, however, a more subtle possibility which is related to information theory. [The biosynthesis of proteins involves the transfer of genetic information from the cell nucleus to the ribosomes (chapter 5).] Every communication device is subject to 'noise' and if a message is transmitted, which contains no redundant symbols, the loss or mutilation of a single symbol by 'noise' destroys the sense of the whole message. For example, the loss or introduction of a single pulse in a digital computer will give rise to errors in computation or may even stop the machine if an instruction in the programme is affected. Likewise, if every amino acid residue in a protein were essential for the manifestation of biological activity, a single genetic mutation, which may be regarded as biological 'noise', would probably be lethal and further evolution would be impossible. By incorporating in the protein molecule a number of amino acids which are redundant, there is a reasonable chance that these will be the ones affected by mutations and the organism and species will survive. In addition to redundant amino acids, proteins may contain amino acids which are degenerate; that is, they can be replaced by a similar amino acid, but not removed completely, without loss of activity. Their function may be to help to maintain the overall shape of the molecule so that the essential amino acids are held in the correct spatial relationship to one another. As explained in chapter 5, the structure of the genetic code is such that some mutations result in the replacement of an amino acid by a structurally similar residue. This possibility together with the degeneracy of the genetic code itself are further natural defences against lethal mutations.

There are three general methods for studying the relationship between the structures of peptides and proteins and their biological activities. First, the structure can be modified by altering the chemical nature of functional groups in the side chains of amino acids or the amino acids can be removed completely by, for example, the use of proteolytic enzymes. Secondly, and this may be regarded as the organic chemist's extension of the first method, analogues can be synthesized in which one or more amino acids are changed at a time. The organic chemist may also adopt the opposite approach of starting with a simple model system and introducing structural complexities in an attempt to increase activity. In each case, the effect of structural modification on biological activity is assayed. Thirdly, by determining the primary structures of homologous proteins from different species and comparing the substitutions, additions and deletions which have occurred as a result of evolutionary trends, it is sometimes possible to decide that certain amino acids are either redundant or degenerate. In the special case of enzymes, kinetic studies can often provide some evidence about which amino acids are involved in the catalytic processes for which they are responsible.

When chemical or enzymic modification of an enzyme has been carried out, it is important to determine not only which amino acids have been affected, but also if these form part of the catalytic site or the binding state, if they help to maintain the correct molecular conformation or if they are apparently redundant. If the presence of a specific substrate or competitive inhibitor does not influence the rate or extent of modification and if the modified enzyme is catalytically indistinguishable from the parent enzyme, the amino acids affected by the reagent cannot form part of the active centre. This does not necessarily mean that the affected groups are redundant; they may assist in maintaining the overall conformation of the enzyme molecule. If, however, an amino acid can be removed by treatment with carboxypeptidase, for example, and if the catalytic activity and enzyme stability are unaffected, then clearly it is functionally redundant.

If modification results in complete loss of enzymic activity and if the presence of a specific substrate or competitive inhibitor offers some protection against the modifying reagent, the active centre must be affected. It is always necessary to carry out control experiments to ensure that physical conditions such as temperature and pH, which are used for the modifying reaction, do not of themselves produce irreversible denaturation. If the specificity of the modifying reagent is low and several sites in the enzyme molecule are altered, the results are much more difficult to interpret. Recently, some highly selective enzyme inhibitors have been discovered. They have a structure which closely resembles that of a good substrate or competitive inhibitor of the particular enzyme; the inhibitor is thus guided to the active centre and is initially *reversibly* bound there. The inhibitor, however, also contains some reactive group; the latter forms a covalent bond with a suitable functional group near or in the active centre while the inhibitor is reversibly bound. The enzyme thus becomes *irreversibly* inhibited and modified. This type of inhibitor might be loosely described as an 'anti-enzyme guided missile'. Some idea of the selectivity attainable by the use of this type of compound may be derived from some work on lactate dehydrogenases (Baker, 1964). Salicylic acid is a competitive inhibitor for lactate dehydrogenases. These enzymes probably contain an essential —SH group in the active centre. 4-Iodoacetamidosalicylic acid irreversibly inhibits lactate dehydrogenase of rabbit skeletal muscle, presumably because the salicylic acid portion of the structure is reversibly bound at the active centre and the iodoacetamido group is able to alkylate the essential —SH group. Lactate dehydrogenase from beef heart, on the other hand, is not irreversibly inhibited by 4-iodoacetamidosalicylic acid. Assuming that the beef heart enzyme also contains an essential —SH group, it must then be concluded that the 4-iodoacetamidosalicylic acid is bound at the active centre in such a way that reaction between the —SH group and the iodoacetamido group is sterically impossible. This marked difference in behaviour of the inhibitor towards the same type of enzyme from different species and tissues signposts a new approach to the design of drugs. If, for example, an enzyme suffered a change of amino acid sequence and molecular conformation when a cell became malignant, it might be possible to synthesize a compound which would selectively and irreversibly inhibit the enzyme in the cancerous tissue without affecting the normal enzyme. If the enzyme were essential for the metabolism or preferably the replication of the malignant cell, the inhibitor might be valuable in the chemotherapy of cancer.

When modification produces a partially active enzyme, several interpretations are possible. If the presence of a specific substrate or competitive inhibitor offers some protection, it is reasonable to conclude that the active centre is implicated, but the extent of reaction at the catalytic site cannot be determined without further information. For example, if the modification of a single amino acid residue in the enzyme molecule lowers the activity to 10 per cent of that of the original enzyme, the modified enzyme could be homogeneous and have a catalytic efficiency of 10 per cent of the original enzyme or it could be a mixture comprising 10 per cent of unmodified enzyme and 90 per cent of completely inactivated enzyme. Again, if more than one site in the molecule is modified, the results are even less definitive, since the product could consist of a mixture of several partially active derivatives together with completely inactive and completely active enzyme. For enzymes which form a *covalent* enzyme-substrate intermediate, these various possibilities can be distinguished in favourable cases by means of an 'all-or-none' assay of the number of catalytic centres which are still functional, albeit with impaired efficiency.

The principle of the 'all-or-none' assay can be illustrated with the enzyme phosphoglucomutase, which catalyses the interconversion of glucose-1-phosphate and glucose-6-phosphate through the intermediate, glucose-1,6-diphosphate. The enzyme is alternately phosphorylated and dephosphorylated during the reaction:

Glucose-1-phosphate + Phospho-enzyme	⇌	Glucose-1,6-diphosphate + Enzyme	⇌	Glucose-6-phosphate + Phospho-enzyme

The phospho-enzyme, labelled with ^{32}P-phosphate, is incubated with a large excess of unlabelled glucose-1-phosphate or glucose-6-phosphate for a few minutes. Enzyme, which still retains any activity, even if its catalytic efficiency is badly impaired, loses its isotopic label during this time. After precipitation of the enzyme with trichloroacetic acid at the end of the incubation, the supernatant is assayed for radioactivity. The ratio of the amount of ^{32}P in the supernatant to that in the original ^{32}P-phospho-enzyme represents the fraction of enzyme which is catalytically active.

6.2. Methods for modifying protein structures. There are three broad approaches for modifying the structure of a protein; amino acid residues can be added to or removed from selected points or the side chains of particular amino acids can be structurally altered. Whichever approach is used, there are some general points to be considered. First, it is important to determine the extent of the structural modification. Analysis of hydrolysates for amino acids is an obvious technique. Where new groups are introduced by substitution, the use of radioactive isotopes or groups with characteristic spectra are also convenient. Secondly, it is necessary to determine where modification has occurred. If the primary structure of the protein is unknown, it may only be possible, for example, to show that a particular type of amino acid has been affected. If the modification involves the introduction of, for example, a radioactive group, it should be possible to determine the sequence of amino acids around the point or points of modification. Thirdly, the results of modification experiments which eliminate biological activity are more significant if the modification can be reversed with concomitant regeneration of biological activity. Finally, as mentioned above, the method of modification should be as selective as possible.

The addition of amino acid residues to an existing protein can be accomplished using oxazolid-2,5-diones (see p. 84). This procedure has been used to investigate which amino acids impart the property of antigenicity to proteins. For example, gelatin is weakly antigenic but becomes highly antigenic when about 2 per cent of tyrosine is introduced as polytyrosyl chains by the reaction between 4-(*p*-hydroxybenzyl)oxazolid-2,5-dione and the free amino groups in gelatin. Interestingly, incorporation of polytyrosyl side chains into trypsin gives a product which has high enzymic activity but which is more stable to self-hydrolysis than native trypsin.

The removal of amino acid residues from proteins can be effected by proteolytic enzymes such as aminopeptidase, carboxypeptidases, trypsin etc. These are discussed in chapter 2 and it is only necessary here to stress that misleading results are all too easily obtained if the protein to be degraded and the proteolytic enzyme are not pure. For example, degradation of the mercury derivative of the proteolytic enzyme, papain, by an impure preparation of aminopeptidase led to the erroneous view that approximately two-thirds of the molecule of almost 200 amino acid residues could be removed from the *N*-terminal end without loss of activity. In fact, more recent work has shown that the —SH group of a cysteine residue near to the *N*-terminus is part of the active centre.

There are numerous chemical procedures for modifying the structure of a protein. Acylation may occur at nucleophilic sites such as amino, thiol, and hydroxyl groups. Reagents such as acid anhydrides are unselective and generally acylate all accessible nucleophilic groups. After acylation, ϵ-amino groups, which are normally protonated at neutral pH, become uncharged or even negatively charged if succinic anhydride is used. This type of structural modification thus provides a useful means of investigating the relationship between the state of charge of a protein and its biological activity. *N*-Acetylimidazole is a more selective reagent, since it acetylates the hydroxyl groups of tyrosyl residues preferentially and then only when these are 'free'. The progress of *O*-acetylation of tyrosine residues can be followed spectrophotometrically. Proteins can also be carbamoylated (§ **2.5**) and, by carefully controlling the pH at about 6–7, reaction can be limited in the main to α-amino groups. ϵ-Amino groups, because they have higher pK_a values react at higher pH values. Tyrosine hydroxyl groups may also react, but the resultant carbamate esters are generally unstable at alkaline pH values. A serine hydroxyl group at the active centre of certain hydrolytic enzymes such as chymotrypsin, trypsin, cholinesterases and other esterases is unusually easily acylated or carbamoylated with loss of enzymic activity, although the substituent is lost again at alkaline pH values. The same hydroxyl group is readily phosphorylated by reagents such as di-isopropyl phosphorofluoridate, $[(CH_3)_2CH.O]_2P(:O).F$ (see p. 134), although reactivation is more difficult or even impossible in this case.

Photo-oxidation in the presence of methylene blue can be used to oxidise the imidazole ring of histidine, the indole ring of tryptophan, tyrosine and the sulphur-containing amino acids. The extent of oxidation can sometimes be controlled by adjustment of pH; histidine is more sensitive in neutral or slightly alkaline solution than it is in acid.

Amino groups are substituted readily with reagents of the general type: $R.C(:NH).OCH_3$ (CXLVIII). *O*-Methyl isourea gives guanidino derivatives (CXLIX: $R = NH_2$), methyl acetimidate gives amidino derivatives (CXLIX; $R = CH_3$) and *O*-methyl-*N*-nitroisourea gives *N*-nitroguanidino derivatives ($R = NO_2.NH$). In the first two cases, the resultant groups are sufficiently basic to retain a proton so that the state of charge is not altered, but the *N*-nitroguanidino group is only feebly basic and does not exist in the protonated state at biological pH values. The *N*-nitroguanidino group is also distinctive, since it absorbs light strongly at 270 mμ; the degree of substitution can hence be readily determined.

$$\underset{\text{(CXLVIII)}}{R.C(:NH).OCH_3} + R'.NH_2 \rightarrow \underset{\text{(CXLIX)}}{R.C(:NH).NH.R'}$$

Iodination can be carried out under mild conditions by reagents such as hypoiodite, iodine-potassium iodide or iodine monochloride. Depending on conditions, accessible tyrosine residues are converted into mono- and/or di-iodotyrosine. Iodine monochloride is the preferred reagent since the other methods often lead to iodination of histidine residues in addition to tyrosine. Tyrosine can also be nitrated in the position *ortho* to the hydroxyl group by tetranitromethane at room temperature and at slightly alkaline pH values.

Unfortunately, there is no mild method for modifying the alkyl side chains of alanine, valine, leucine or isoleucine. Analogues of small, biologically active peptides have been synthesized, however, which contain different alkyl side chains from the natural peptide and there is no doubt that such groups play a part in determining biological activity.

6.3. Oxytocin and vasopressin. The posterior pituitary gland contains oxytocin (CL), which causes contraction of uterine muscle, increases pressure in lactating mammary gland and decreases the blood pressure in chickens, and vasopressin (CLI), which increases blood pressure and inhibits diuresis in the rat. Some non-mammalian vertebrates contain a hybrid molecule, vasotocin (CLII) and other variations have been found.

$$\overline{\text{CyS-Tyr-Ile-Glu(NH}_2\text{)-Asp(NH}_2\text{)-CyS}}\text{-Pro-Leu-Gly-NH}_2$$

(CL)

$$\overline{\text{CyS-Tyr-Phe-Glu(NH}_2\text{)-Asp(NH}_2\text{)-CyS}}\text{-Pro-X-Gly-NH}_2$$

[CLI; X = Lys (pig) or Arg (other mammals)]

$$\overline{\text{CyS-Tyr-Ile-Glu(NH}_2\text{)-Asp(NH}_2\text{)-CyS}}\text{-Pro-Arg-Gly-NH}_2$$

(CLII)

In view of the similarity between oxytocin and vasopressin, it is not surprising that the activities of one are manifested by the other in some degree. Vasotocin possesses high levels of both oxytocic and pressor activities. Oxytocin was synthesized by du Vigneaud in 1954 and since then many analogues have been synthesized (Schröder & Lübke, 1966) in attempts to determine which amino acids are essential for activity. Vasotocin was synthesized as an analogue of oxytocin and vasopressin before it had been found to occur naturally. Some synthetic analogues show more biological specificity than the natural hormones. For example, Phe^2-Lys^8-vasopressin (i.e. CLI in which X is Lys and Phe replaces Tyr) has about one-fifth of the pressor activity of lysine vasopressin, but its antidiuretic and oxytocic activities are decreased by a much larger factor.

The activity of oxytocin soon disappears *in vivo*, probably as a result of hydrolysis by an enzyme in blood, oxytocinase, which appears to hydrolyse the $\overline{\text{CyS-Tyr}}$ peptide bond. Some synthetic analogues have a more protracted or a higher activity than the natural hormones. One of these, desamino-oxytocin, has β-mercaptopropionic acid in place of the *N*-terminal CyS- residue; it is resistant to oxytocinase, possibly because it has no α-amino group.

6.4. Adrenocorticotropic and melanocyte-stimulating hormones. The anterior pituitary gland contains adrenocorticotropic hormone (also referred to as ACTH or corticotropin) (CLIII), which stimulates the adrenal cortex to produce corticosteroids, has lipolytic activity in adipose tissue, and causes hyperglycaemia.

1 2 3 4 5 6 7 8 9 10 11 12 13 14 15 16 17 18
Ser-Tyr-Ser-Met-Glu-His-Phe-Arg-Trp-Gly-Lys-Pro-Val-Gly-Lys-Lys-Arg-Arg-

19 20 21 22 23 24 25 26 27 28 29 30 31 32 33 34
Pro-Val-Lys-Val-Tyr-Pro-Asp-Gly-Ala-Glu-Asp-Glu(NH_2)-Leu-Ala-Glu-Ala-

35 36 37 38 39
Phe-Pro-Leu-Glu-Phe (CLIII)

A number of fragments and analogues of fragments as well as the complete hormone from pig (CLIII) have been synthesized. Peptides comprising the first thirteen to sixteen amino acids from the *N*-terminus have very low activity; a peptide amide containing the first eighteen residues has high activity, and peptides containing the first twenty-three or twenty-four amino acids are as active as the natural hormone. The last fifteen residues thus appear to be redundant and it is interesting that variations in sequence, found in adrenocorticotropic hormone from different species, occur between residues twenty-five to thirty-three inclusive (CLIV–CLVII).

	25 26 27 28 29 30	31 32 33	
Pig	-Asp-Gly-Ala-Glu-Asp-Glu(NH_2)	-Leu-Ala-Glu-	(CLIV)
Cow	-Asp-Gly-Glu-Ala-Glu-Asp	-Ser-Ala-Glu(NH_2)-	(CLV)
Sheep	-Ala-Gly-Glu-Asp-Asp-Glu	-Ala-Ser-Glu(NH_2)-	(CLVI)
Man	-Asp-Ala-Gly-Glu-Asp-Glu(NH_2)	-Ser-Ala-Glu-	(CLVII)

Oxidation of methionine to the sulphoxide at position 4 in adrenocorticotropic hormone markedly lowers the activity and it seemed reasonable to suppose that this amino acid was essential for activity. An eicosapeptide amide has been synthesized, however, in which α-aminobutyric acid replaced Met^4 in the sequence 1–20 of the hormone, and it had high activity. α-Aminobutyric acid and methionine are isosteres and perhaps the side-chain helps in binding the hormone to the receptor; oxidation of the methionine to the highly polar sulphoxide would interfere with the formation of a hydrophobic bond. It must be concluded that methionine has an auxiliary role but is not essential.

α-MSH	CH_3.CO-Ser-Tyr-Ser-Met-Glu-His-Phe-Arg-Trp-Gly-Lys-Pro-Val-NH_2
β-MSH (horse)	Asp-Glu-Gly-Pro-Tyr-Lys-Met-Glu-His-Phe Arg-Trp-Gly-Ser-Pro-Arg-Lys-Asp
β-MSH (cow)	Asp-Ser-Gly-Pro-Tyr-Lys-Met-Glu-His-Phe-Arg-Trp-Gly-Ser-Pro-Pro-Lys-Asp
β-MSH (pig)	Asp-Glu-Gly-Pro-Tyr-Lys-Met-Glu-His-Phe-Arg-Trp-Gly-Ser-Pro-Pro-Lys-Asp
β-MSH (monkey)	Asp-Glu-Gly-Pro-Tyr-Arg-Met-Glu-His-Phe Arg-Trp-Gly-Ser-Pro-Pro-Lys-Asp
β-MSH (man)	Ala-Glu-Lys-Lys-Asp-Glu-Gly-Pro-Tyr-Arg-Met-Glu-His-Phe-Arg-Trp-Gly-Ser-Pro-Pro-Lys-Asp

Fig. 6.1. Melanocyte-stimulating hormones.

Melanocyte-stimulating hormones (MSH), present in the intermediate lobe of the pituitary, stimulate cells called melanocytes to produce the black pigment melanin. Most species examined appear to produce two melanocyte-stimulating hormones. α-MSH (fig. 6.1), present in sheep, horse, cow, monkey and pig, is unusual, since the α-amino group is acetylated and the *C*-terminal valine is in the form of its amide. Sequence differences are found in β-MSH from different species and the hormone from man is the most complex so far found (fig. 6.1). The most remarkable feature of the structure of the melanocyte-stimulating hormones is the occurrence of the sequence: -Met-Glu-His-Phe-Arg-Trp-Gly- in each and also in adrenocorticotropic hormone and it is significant that the latter has weak melanocyte-stimulating activity. The synthetic pentapeptide His-Phe-Arg-Trp-Gly is weakly active and extension of the sequence at either end gradually increases activity. Surprisingly, the diastereoisomeric pentapeptide containing D-phenylalanine is more active than the all-L-peptide, while the corresponding all-D-peptide inhibits the activity of the all-L-peptide.

The presence of an *N*-acetyl group has an important influence on the activity of α-MSH. The unprotected tridecapeptide has about 7 per cent and an unprotected undecapeptide, containing residues 3–13, has only 1 per cent of the activity of the natural hormone. The *N*-acetylated undecapeptide, however, has 25 per cent of the activity of α-MSH. It is possible that the *N*-acetyl group protects the hormone from proteolytic enzymes or, alternatively, the absence of a protonated group at the *N*-terminus improves binding at the receptor.

6.5. Gastrin. Gastrin is a peptide hormone isolated from gastric antral mucosa. It has a complex pattern of biological activity, but, under suitable conditions, it stimulates the secretion of gastric acid and pepsin and increases gastric tone and motility. Two peptides were obtained from pigs; gastrin I was shown by degradation and chemical synthesis to have the structure (CLVIII)

Glu*-Gly-Pro-Trp-Met†-Glu-Glu-Glu-Glu-Glu-Ala-Tyr‡-Gly-Trp-Met-Asp-Phe-NH_2 (CLVIII)

* Present as a 'pyroglutamyl' residue, $\underset{\text{NH}—\text{CH}—\text{CO}—}{\overset{\text{CO}—(\text{CH}_2)_2}{|\quad\quad|}}$

† In man, Met is replaced by Leu and in sheep by Val.

‡ In gastrin II, the tyrosine residue is present as the *O*-sulphate ester.

The *C*-terminal tetrapeptide amide, Trp-Met-Asp-Phe-NH_2, was synthesized and shown to possess almost the complete range of gastrin-like activities, so that much of the remaining structure appears to be redundant. Biological testing of a number of analogues of the *C*-terminal tetrapeptide amide indicates that (*a*) the indole ring of tryptophan is probably important for binding to the receptor; (*b*) methionine can be exchanged for other amino acids with hydrophobic side chains, but oxidation to the sulphoxide or sulphone causes extensive loss of activity; (*c*) aspartic acid seems to be essential; (*d*) replacement of phenylalanine by tyrosine causes almost complete loss of activity, but the analogue containing *O*-methyltyrosine is quite active; (*e*) the terminal amide appears to be essential. It is interesting to note that three of the four amino acids in this tetrapeptide occur in the heptapeptide sequence, which is common to adrenocorticotropic hormone and the melanocyte-stimulating hormones.

6.6. Bradykinin and kallidin. Treatment of a plasma globulin fraction with certain snake venoms or trypsin affords a nonapeptide, bradykinin (CLIX):

1 2 3 4 5 6 7 8 9
Arg-Pro-Pro-Gly-Phe-Ser-Pro-Phe-Arg (CLIX)

The related decapeptide, kallidin, is formed when the enzyme, kallikrein, from submaxillary gland or other sources is used instead of snake venom or trypsin. Kallidin differs from bradykinin in having an extra amino acid, lysine, at the *N*-terminus. Bradykinin is a powerful vasodilator, stimulates smooth muscle and causes pain; it may be responsible in part for the inflammatory response.

A large number of analogues of bradykinin have been synthesized (Schröder & Lübke, 1966) with a view to elucidating which amino acids are important in determining biological activity. It appears that both arginine residues are required and they must have the L-configuration. If Pro^2 or Pro^7 are replaced by alanine or omitted, serious or complete loss of activity ensues. In contrast Pro^3 can be replaced by alanine with little loss in activity. Replacement of Gly^4 even by alanine produces a large drop in reactivity. On the other hand, Ser^6 can be replaced by one of several amino acids without serious loss of activity. Phe^8 can have a *p*-fluoro-substituent or it can have the D-configuration, but replacement of this or Phe^5 by alanine leads to a considerable loss of activity.

6.7. Eledoisin and physalaemin. Eledoisin (CLX), isolated from the posterior pituitary glands of *Eledone moschata* (a Mediterranean cephalapod), and physalaemin (CLXI), isolated from the skin of *Physalaemus fuscumaculatus* (a South American amphibian), are structurally related peptides which stimulate smooth muscle and have powerful vasodilator and hypotensive actions. Although their pharmacological properties qualitatively resemble those of bradykinin, their chemical structures are quite distinct.

```
1     2   3   4   5          6   7   8   9   10  11
Glu* Pro-Ser-Lys-Asp       -Ala-Phe-Ile-Gly-Leu-Met-NH2          (CLX)
Glu*-Ala-Asp-Pro-Asp(NH2)-Lys-Phe-Tyr-Gly-Leu-Met-NH2           (CLXI)
```

* Present as a 'pyroglutamyl' residue,

```
CO—(CH2)2
|      |
NH—CH—CO—
```

Over 150 fragments and analogues of eledoisin have been synthesized and the following relationships between structure and biological activity have been revealed: (*a*) the pentapeptide amide representing the *C*-terminal sequence of eledoisin has detectable activity and this increases as the chain is lengthened towards the

N-terminus, reaching a maximum in the nona- and deca-peptides, which are more active than eledoisin itself; (*b*) varying the terminal amide group profoundly lowers activity; (*c*) omission or replacement of methionine profoundly lowers activity; (*d*) although the *C*-terminal portion of eledoisin is important, a certain amount of variation in the amino acids at positions 8 and 9 is possible without serious loss of activity.

6.8. Angiotensin (Hypertensin). The action of a kidney enzyme, renin (not to be confused with another proteolytic enzyme, rennin, found in calf stomach), on a plasma α_2-globulin liberates a decapeptide, angiotensin I. A second proteolytic enzyme in the blood splits off the dipeptide, His-Leu, from the *C*-terminus of angiotensin I and leaves angiotensin II (CLXII):

```
1   2   3   4   5   6   7   8
Asp-Arg-Val-Tyr-Ile*-His-Pro-Phe   (CLXII)   (horse and pig)
```

* Valine in cow angiotensin II.

Angiotensin increases blood pressure by causing vascular contraction and helps to control salt balance by releasing aldosterone.

The synthesis and biological assay of numerous analogues has thrown some light on the relationship between structure and activity. The *C*-terminal phenylalanine seems to be essential; omission or replacement by L-alanine or D-phenylalanine destroys activity and even conversion of the free acid into an ester or an amide diminishes activity to a low level. Pro^7 also seems to be essential, since replacement by alanine gives an inactive peptide. At the other terminus, a variety of modifications to the aspartic acid residue or even omission does not profoundly affect activity.

The importance of the *C*-terminal sequence in many biologically active peptides has been stressed by Stewart & Woolley (1965), and this importance is brought out by the biological activity of fragments and analogues of gastrin, eledoisin, angiotensin and other peptides. On the other hand, it is the *N*-terminal sequence of adrenocorticotropic hormone which determines the activity. It is probably unwise to try to generalize; any redundancy in structure might depend on the biological conditions under which the peptide exists. If exposure to aminopeptidases is likely to occur, then it would not be surprising to find that protection has been acquired during evolution by the addition of redundant amino acids at the *N*-terminus. On the other hand, if exposure to carboxypeptidases is more likely, redundant residues would be expected to occur at the *C*-terminus. It should also be pointed out that a small peptide *per se* is probably inactive. Klaus Hofmann (see p. 133) has suggested that these small peptides are bound to specific proteins (receptors) to form active enzymes. If this is the case, the *C*-terminus of a small peptide such as gastrin, for example, could be bound at the *N*-terminus of its receptor(s) to form one or more enzymes which evoke the physiological responses associated with the action of gastrin. There seems to be no reason, however, why the converse situation should not hold with other peptides such as adrenocorticotropic hormone. Indeed, it is possible to imagine non-covalent bonding occurring between peptide and protein at any point along their length if the protein conformation is favourable. The mechanism of action of peptide hormones (and perhaps all hormones) may then be regarded as an extension of the relation between an enzyme and its coenzyme.

6.9. Insulin. Although insulin (fig. 6.2) was the first protein whose sequence of amino acids was determined (Sanger, 1955), little is known of the relationship between its structure and biological activity. Fission of the disulphide bridges leads to two fragments which are inactive. Removal of *C*-terminal alanine from the *B*-chain is without much effect, but if the *C*-terminal octapeptide is removed by fission of the Arg-Gly peptide bond with trypsin, the product is practically inactive. The *C*-terminal asparagine of the *A*-chain appears to be important for manifestation of biological activity. Photo-oxidation of insulin, which destroys both histidine residues, also destroys activity. The amino groups may be acetylated without affecting biological activity, but esterification of carboxyl groups inactivates the hormone. Although it appears that the disulphide linkage between CyS^6 and CyS^{11} in the *A*-chain is important, the sequence of amino acids inside the ring need not be unique. Indeed, species variations are found here. Bovine insulin, which has the sequence $\overset{8}{\text{Ala}}\text{-}\overset{9}{\text{Ser}}\text{-}\overset{10}{\text{Val}}$, is active in man, while human insulin has the sequence $\overset{8}{\text{Thr}}\text{-}\overset{9}{\text{Ser}}\text{-}\overset{10}{\text{Ile}}$.

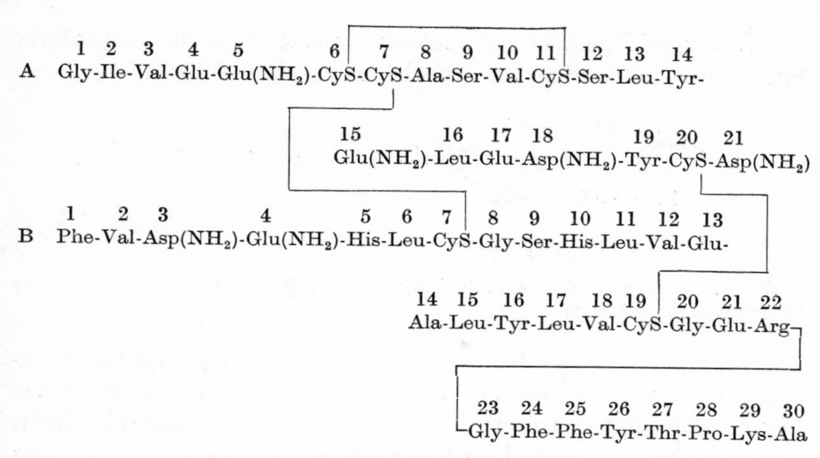

Fig. 6.2. Structure of bovine insulin.

6.10. Peptide antibiotics. The structures of many peptide antibiotics have been clearly established, but the relationship between these and antibacterial activity is much less well understood. There are a number of recurrent structural features which could conceivably contribute to antibiotic behaviour: (*a*) peptide antibiotics usually contain amino acids which are not normally found in proteins, (*b*) they frequently contain one or more amino acids with the D-configuration, and (*c*) they are frequently cyclic peptides.

Gramicidin S (CLXIII) contains ornithine, which does not normally occur in proteins, it contains D-phenylalanine, and is a cyclic decapeptide.

Val→Orn→Leu→D-Phe→Pro
↑ ↓
Pro←D-Phe←Leu←Orn←Val

(CLXIII)

Val→Orn→Leu
↑ ↓
Pro← D-Phe

(CLXIV)

The cyclic pentapeptide (CLXIV) and some analogous linear decapeptides have been synthesized and were found to be inactive. This suggests that the cyclic structure and hence the molecular conformation are important; the presence of an unusual amino acid or a D-amino acid *per se* is not sufficient for the manifestation of antibacterial properties. Moreover, ornithine can be replaced by lysine in the cyclodecapeptide without loss of activity.

Polymyxin B_1 (CLXV) is another peptide antibiotic containing ten amino acid residues, one of which is D-phenylalanine.

```
Dab ──→ Dab ───→ Thr
 ↑                  ↘ γ
 │                     α
 │                    Dab ←─ Dab ←─ Thr ←─ Dab ←─ Ipel
 │                  ↙
Leu ←─ D-Phe ←── Dab
```

(CLXV)

(Dab is a residue of $\alpha\gamma$-diaminobutyric acid; Ipel is an isopelargonoyl or 6-methyloctanoyl group)

Polymyxin B_1 differs from gramicidin S, since it contains $\alpha\gamma$-diaminobutyric acid, the lower homologue of ornithine, and it has a cycloheptapeptide ring with an *N*-acylated tripeptide side chain. The side chain is located on the α-amino group of the branching molecule of $\alpha\gamma$-diaminobutyric acid, while the ring is closed with an amide bond between the carbonyl group of a threonyl residue and the γ-amino group of $\alpha\gamma$-diaminobutyric acid. The γ-amino groups of the remaining residues of $\alpha\gamma$-diaminobutyric acid are free. It was originally thought that the *N*-terminal residue of $\alpha\gamma$-diaminobutyric acid in the side chain had the D-configuration. This diastereoisomer of polymyxin B_1 was synthesized by Vogler and shown to have approximately the same activity as the natural antibiotic, an observation which supports the view that the presence of D-amino acid itself is not sufficient to account for the biological activity. It is much more likely that the overall molecular conformation and the general chemical nature (hydrophobic or hydrophilic) of the side chains of the constituent amino acids are the chief features which give rise to antibacterial activity and it is possible that Hofmann's views of the method of action of hormones could be extended to include peptide antibiotics if the latter form stable complexes with essential enzymes. It is perhaps not without importance that small cyclic peptides and peptides containing D-amino acids are inclined to be resistant to proteolytic enzymes. The metabolism of the antibiotics is thus likely to be fairly slow.

One particularly interesting demonstration of the importance of the overall conformation has been described in connection with the depsipeptide antibiotic enniatin B (CLXVI; $R = (CH_3)_2CH$, $R' = CH_3$). Enantio-enniatin B (CLXVII; $R = (CH_3)_2CH$, $R' = CH_3$), in which the configuration of all six residues is inverted, has been synthesized and has been shown to be as active as the natural antibiotic.

```
          (D)          (L)                               (L)          (D)
          CHR.CO.NR'.CHR                                 CHR.CO.NR'.CHR
         /              \                               /              \
        O                CO                            O                CO
       /                  \                           /                  \
      CO                   O                         CO                   O
     /                      \                       /                      \
(L) CHR                      CHR (D)           (D) CHR                      CHR (L)
     \                      /                       \                      /
      NR'                  CO                        NR'                  CO
       \                  /                           \                  /
        CO               NHR'                          CO               NHR'
         \              /                               \              /
          CHR.O.CO---CHR                                 CHR.O.CO---CHR
          (D)          (L)                               (L)          (D)

             (CLXVI)                                        (CLXVII)
```

It can be seen from the formulae (CLXVI, CLXVII) that rotation of one of the structures by 60° in the plane of the molecule gives two structures of identical stereochemistry with the exception that —CO.NCH_3— and —CO—O— groups are interchanged.

6.11. Ribonuclease. Bovine pancreatic ribonuclease is much more complex in structure than any of the polypeptides discussed so far, but we are perhaps nearer to understanding the mechanism of its action than for any other protein with the possible exception of chymotrypsin and trypsin. The reason is obvious; ribonuclease is an enzyme, whose primary structure has been established (fig. 6.3), and the effect of structural modifications can be assessed by kinetic studies. Indeed, kinetic studies with the intact enzyme have thrown much light on its mechanism of action.

Bovine pancreatic ribonuclease hydrolyses ribonucleic acids to a mixture of pyrimidine nucleotides and oligonucleotides which terminate in a pyrimidine nucleotide. Cyclic uridine- and cytidine-2′,3′-phosphates are intermediates in the reaction and can be used as simple substrates for the enzyme. Kinetic studies have indicated that the imidazole rings of two histidine residues form part of the active centre of ribonuclease (Mathias, Deavin & Rabin, 1964); it is likely that one imidazole is unprotonated while the other is in

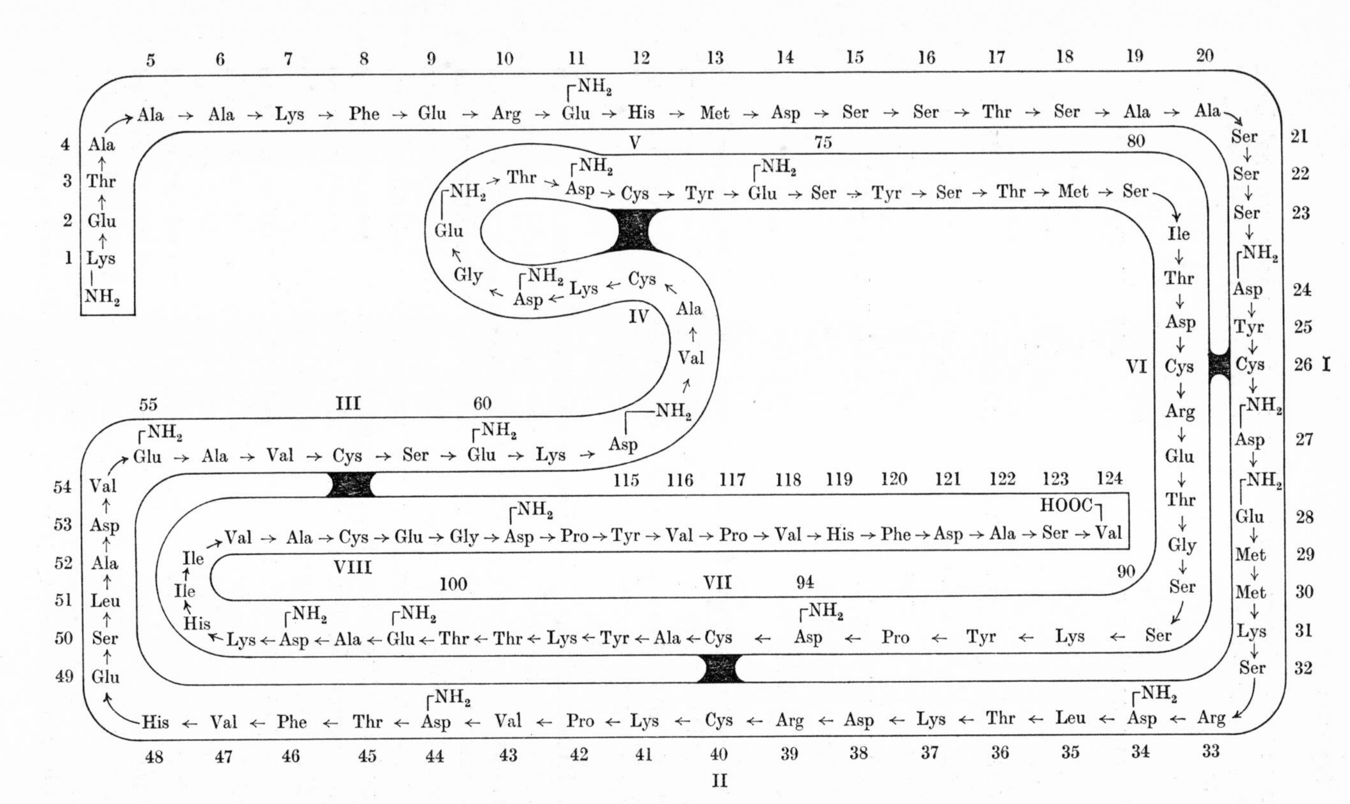

Fig. 6.3. Primary structure of bovine pancreatic ribonuclease. [D. G. Smyth, W. H. Stein & S. Moore (1963). *J. biol. Chem.* **238**, 227.]

the form of the conjugate acid. The inactivation of ribonuclease by photo-oxidation also indicates that histidine is essential for catalytic activity. Since ribonuclease contains four histidine residues it is essential to know which two are the more important. Alkylation of ribonuclease under appropriate conditions with iodoacetic acid gave rise to two distinct monocarboxymethyl derivatives. One of these was almost inactive and sequence analysis revealed that His[119] had been alkylated at N_1 (CLXVIII). The other monocarboxymethyl derivative had 5–7 per cent of the activity of the native enzyme and, in this case, His[12] was shown to be alkylated at N_3 (CLXIX).

```
              |
              NH
              |
HC=====C—CH—CO—
|      |
N      N.CH2.CO.OH
 \\   /
   CH

     (CLXVIII)
```

```
                         |
                         NH
                         |
              HC=====C—CH—CO—
              |      |
HO.CO.CH2.N      N
               \   //
                 CH

             (CLXIX)
```

Alkylation of one histidine residue apparently blocks the alkylation of the other and it must be concluded that, although the two histidine residues are at opposite ends of the amino-acid sequence, they must be in juxtaposition as a result of the folding of the polypeptide chain.

Cleavage of disulphide linkages by reduction inactivates ribonuclease, but it seems probable that the interactions between the side chains of amino acids maintain a secondary and tertiary structure which is similar to that of the native enzyme, since aerial oxidation of the reduced enzyme regenerates a high level of enzyme activity.

When ribonuclease is exposed briefly to the action of a proteolytic enzyme from *B. subtilis*, subtilisin, a single peptide bond between $\overset{20}{\text{Ala}}$-$\overset{21}{\text{Ser}}$ is cleaved. The peptide comprising the first twenty amino acids (S-peptide) and the remainder of the original molecule (S-protein) are strongly bound to one another at neutral pH and give a single peak when subjected to ion-exchange chromatography. Careful treatment with trichloroacetic acid, however, precipitates the S-protein and leaves the S-peptide in solution.

Separately, the two fragments are inactive, but a mixture of the two is almost as active as the original enzyme. It should be noticed that each fragment contains one of the essential histidine residues. Presumably, a mixture of S-peptide and S-protein form a complex which adopts a conformation similar to that of ribonuclease with the two essential histidine residues in close proximity. A number of peptides related to the S-peptide have been synthesized by Hofmann and his colleagues. An undecapeptide containing the first eleven residues of the S-peptide does not yield any enzymic activity when mixed with S-protein; this is not surprising, since His^{12} is missing. A tridecapeptide, however, containing the first thirteen residues of the S-peptide gave high activity in admixture with S-protein. The ability of the *N*-terminal tetradecapeptide to give even higher activity with S-protein, approximating to that given by S-peptide and S-protein, suggests that Asp^{14} plays an auxiliary role, probably by forming part of a binding site between the peptide and S-protein. Two synthetic dodecapeptides corresponding to the first twelve amino acids of the S-peptide, but in which His^{12} was replaced by either β-(pyrazolyl-1) alanine (CLXX) or β-(pyrazolyl-3)alanine (CLXXI), failed to give enzymic activity when mixed with S-protein.

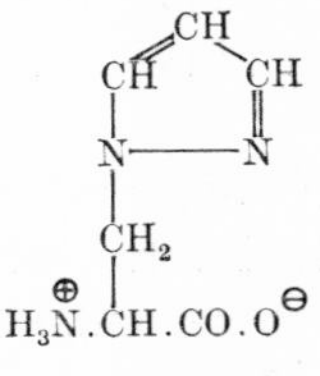

$H_3\overset{\oplus}{N}.CH(CH_2\text{-pyrazol-1-yl}).CO.O^{\ominus}$

(CLXX)

$H_3\overset{\oplus}{N}.CH(CH_2\text{-pyrazol-3-yl}).CO.O^{\ominus}$

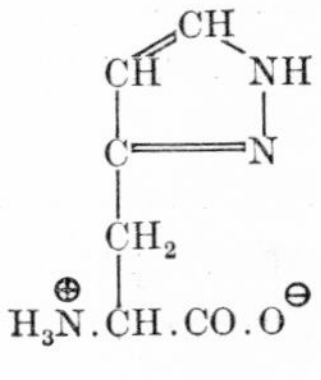

(CLXXI)

Although histidine, β-(pyrazolyl-1)alanine and β-(pyrazolyl-3)-alanine are all isomeric, the last two have much lower pK_a values than histidine; this may be connected with the failure of the two dodecapeptides to generate activity in admixture with S-protein.

From the foregoing results, it appears that the sequence, Ser^{15}-Ser^{16}-Thr^{17}-Ser^{18}-Ala^{19}, is unnecessary for enzymic activity and possibly assists only by maintaining the correct spatial relationship between the essential groups.

As a result of his work on the interaction of synthetic peptides and S-protein, Hofmann has suggested that the peptide hormones may be bound to their receptors in a similar manner to that found with S-peptide and S-protein. The resulting complex would exhibit a specific enzymic activity associated with the biological activity of the hormone. Just as a single coenzyme can be bound to different enzymes, so the peptide hormones might be bound to different receptors and this could explain the multiple activities of certain hormones. So far, there is no direct evidence to support or disprove this hypothesis, but it should stimulate the search for and chemical study of hormone receptors.

During reaction of ribonuclease with 1-fluoro-2,4-dinitrobenzene, complete inactivation occurs when substitution of the ϵ-amino group of Lys^{41} has taken place. This amino group is unusually reactive except in phosphate buffer. The latter observation provides a clue concerning its function. It is normally protonated at neutral pH, of course, and its inertness in phosphate buffer may be due to the binding of anions there. It is probable then that it assists in binding the phosphate group of the substrate. A recent X-ray crystallographic study of ribonuclease, which had been crystallized in the presence of phosphate, shows that the molecule is folded so that the phosphate anion is close to His^{12}, His^{119}, Lys^{7}, Lys^{41} and His^{48} (Kartha, Bello & Harker, 1967).

In a proposed mechanism of ribonuclease-catalysed reactions (fig. 6.4) (Deavin, Mathias & Rabin, 1966), it is suggested that a substrate, such as an ester of cytidine-3′-phosphate, forms a complex with the enzyme by means of electrostatic and hydrogen bond interactions between the phosphate group of the substrate and an ϵ-amino group and an imidazole ring of the enzyme. The pyrimidine ring is bound by a hydrogen bond through a water molecule to one of the imidazole rings. Conformational changes then allow the second imidazole ring to move close enough to form a hydrogen bond with the hydroxyl group at $C_{2'}$ on the ribose moiety of the nucleotide. It is possible that other interactions occur between enzyme and substrate. Formation of the cyclic phosphate and cleavage of the cytidylic ester bond could be achieved as a result of the electronic shifts indicated by the arrows (fig. 6.4*a*). Notice that one imidazole ring is required in its basic form while the other is required as its conjugate acid. The hydrolysis of the cyclic phosphate to cytidine-3′-phosphate requires first that ROH (fig. 6.4*b*) be replaced by a water molecule, a process likely to occur rapidly in an aqueous medium. Formation of the cytidine-3′-phosphate now involves reversing the electronic shifts involved in the formation of the cyclic phosphate (fig. 6.4*b* → *a*) remembering that ROH is now a water molecule.

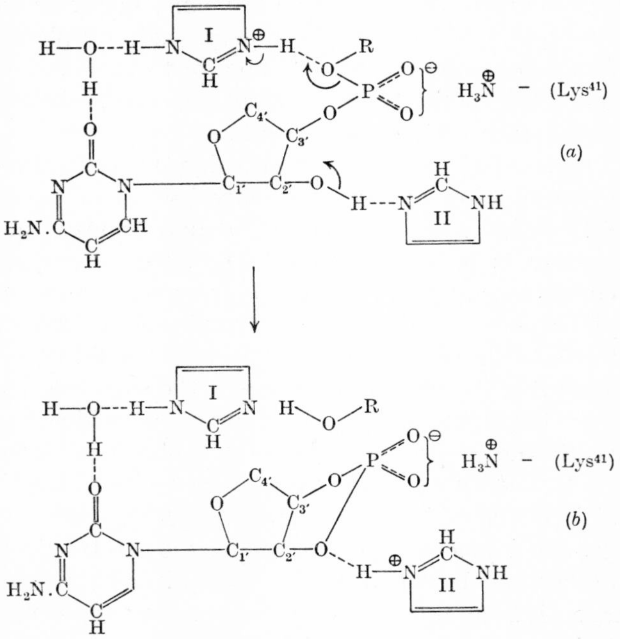

Fig. 6.4. Proposed mechanism for the catalytic action of bovine pancreatic ribonuclease. [After A. Deavin, A. P. Mathias & B. R. Rabin (1966). *Biochem J.* **101**, 14C.]

6.12. α-Chymotrypsin and trypsin. α-Chymotrypsin and trypsin share the property, together with other proteolytic enzymes such as thrombin and elastase and several esterases, of being irreversibly inhibited by di-isopropyl phosphorofluoridate $[(C_3HO_7)_2PO.F]$ and related nerve gases. Total hydrolysis of the inhibited enzymes has revealed that, in each case, the hydroxyl group in the side chain of a residue of serine is phosphorylated. Less drastic hydrolysis of the inhibited enzymes has given phosphorylated peptides and it is clear that the sequences of

amino acids around the reactive seryl residue are remarkably similar (CLXXII–CLXXIX; the seryl residue which is phosphorylated is underlined in each case).

	191 192 193 194 195 196 197 198 199 200 201	
α-Chymotrypsin	-Cys- Met -Gly-Asp-Ser-Gly-Gly-Pro-Leu-Val-CyS-	(CLXXII)
	179 180 181 182 183 184 185 186 187 188 189	
Trypsin	-CyS-Glu(NH_2)-Gly-Asp-Ser-Gly-Gly-Pro-Val-Val-CyS-	(CLXXIII)
Elastase	-CyS-Glu(NH_2)-Gly-Asp-Ser(Gly,Gly,Pro)Leu-His-CyS-	(CLXXIV)
Thrombin	-Asp-Ser-Gly-	(CLXXV)
Cholinesterase	-Phe-Gly-Glu-Ser-Ala-Gly-	(CLXXVI)
Acetylcholinesterase	-Glu-Ser-Ala-	(CLXXVII)
Aliesterase	-Gly-Glu-Ser-Ala-Gly-Gly-	(CLXXVIII)
Alkaline phosphatase	-Thr-Asp-Ser-Ala-Ala-	(CLXXIX)

α-Chymotrypsin and trypsin slowly hydrolyse certain esters, such as *p*-nitrophenyl acetate, which have little structural resemblance to a typical substrate (see p. 36). It was shown spectrophotometrically that there is an initial 'burst' of *p*-nitrophenol before a steady state is established. This suggests that an acylated enzyme is formed as an intermediate in the enzymic reaction:

$$CH_3.CO.O.C_6H_4.NO_2 + \text{Enzyme} \underset{k_{-1}}{\overset{k_1}{\rightleftharpoons}} \begin{array}{c}\text{Enzyme-substrate}\\ \text{complex}\end{array} \xrightarrow{k_2} \text{acetyl-enzyme} + HO.C_6H_4.NO_2$$

$$\text{acetyl-enzyme} \xrightarrow[k_3]{H_2O} \text{Enzyme} + \text{acetic acid}$$

For a 'burst' to be observed, $k_2 > k_3$; aryl esters are more likely to meet this condition than alkyl esters or amides, since they are better acylating agents (see p. 88). Other indirect evidence supports this mechanism. α-Chymotrypsin hydrolyses several esters of *N*-acetyl-L-tryptophan at the same rate, although rate constants for the non-enzymic base-catalysed hydrolysis of the esters cover a wide range. Similar behaviour has been found with trypsin and esters of *N*-benzoyl-L-arginine, *N*-toluene-*p*-sulphonyl-L-arginine, *N*-toluene-*p*-sulphonyl-L-homoarginine and *N*-toluene-*p*-sulphonyl-L-lysine. These observations suggest that the rate-determining step for the hydrolysis of each series of esters is the hydrolysis of the common intermediate, the acylated enzyme. The site of acylation is likely to be the hydroxyl group of the same seryl residue which is phosphorylated when these enzymes are inhibited by di-isopropyl phosphorofluoridate. Indeed, the acetylated enzyme, produced during the hydrolysis of *p*-nitrophenyl acetate by chymotrypsin proved to be sufficiently stable for a sequence analysis to be carried out. The results proved that acetylation and phosphorylation occurred at the same site. It could of course be argued that the mechanism is different with specific substrates and that the presence of the di-isopropylphosphoryl group in the seryl residue merely sterically hinders the approach of a specific substrate. The direct participation of the hydroxyl group in the side chain of serine, however, has been further supported. α-Chymotrypsin can also be totally inhibited by reaction with toluene-*p*-sulphonyl chloride, which reacts at the hydroxyl group of the essential seryl residue. In the presence of base and under mild conditions, β-elimination occurs (CLXXX → CLXXXI) and a dehydro-enzyme is formed which is quite inactive towards specific substrates.

$$CH_3.C_6H_4.SO_2.O.CH_2.\underset{\substack{|\\NH\\|}}{\overset{\substack{|\\CO\\|}}{CH}} \xrightarrow{OH^{\ominus}} CH_3.C_6H_4.SO_3^{\ominus} + CH_2{=}\underset{\substack{|\\NH\\|}}{\overset{\substack{|\\CO\\|}}{C}}$$

(CLXXX) (CLXXXI)

From the effect of pH on the rate of reactions catalysed by α-chymotrypsin and trypsin, it appears that the basic form of an imidazole ring of a histidyl residue is implicated in the catalytic mechanism. Some indirect evidence has been derived from model experiments. Imidazole and its derivatives are nucleophilic catalysts for the hydrolysis of active esters such as *p*-nitrophenyl acetate. To some extent, these results were misleading, since it is now believed that imidazole rings function as a general acid-base catalyst rather than a nucleophilic catalyst. More direct evidence has been obtained using some of the 'anti-enzyme guided missiles' mentioned earlier (p. 115). Thus, L-1-chloro-4-phenyl-3-toluene-*p*-sulphonamido-2-butanone (CLXXXII) and L-7-amino-1-chloro-3-toluene-*p*-sulphonamido-2-heptanone (CLXXXIII) irreversibly inhibit α-chymotrypsin and trypsin respectively. Their structures simulate those of typical substrates such as esters of *N*-toluene-*p*-sulphonyl-L-phenylalanine and -lysine, but they contain no C—O or C—N bond which can be cleaved. Instead, they are powerful alkylating agents. Degradative studies with enzymes which had been inhibited with these reagents revealed that the imidazole ring of a histidyl group (His^{57} in chymotrypsin and His^{46} in trypsin) had been alkylated. Moreover the sequences about the susceptible group are similar for α-chymotrypsin (CLXXXIV) and trypsin (CLXXXV).

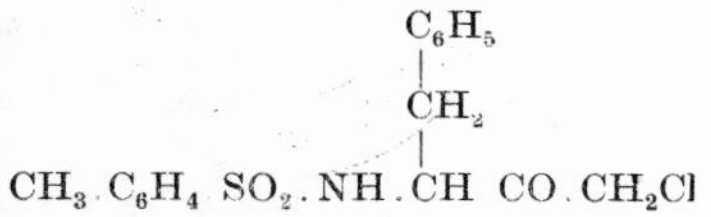

$$\begin{array}{c} C_6H_5 \\ | \\ CH_2 \\ | \\ CH_3.C_6H_4.SO_2.NH.CH.CO.CH_2Cl \end{array} \qquad \begin{array}{c} \overset{\oplus}{N}H_3Cl^{\ominus} \\ | \\ (CH_2)_4 \\ | \\ CH_3.C_6H_4.SO_2.NH.CH.CO.CH_2Cl \end{array}$$

(CLXXXII) 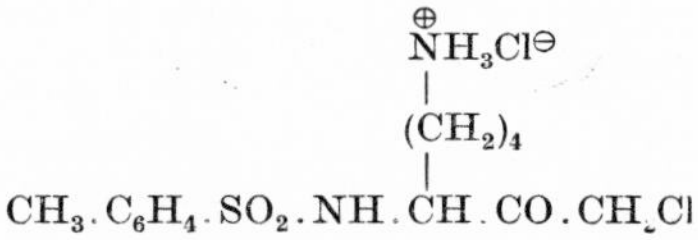 (CLXXXIII)

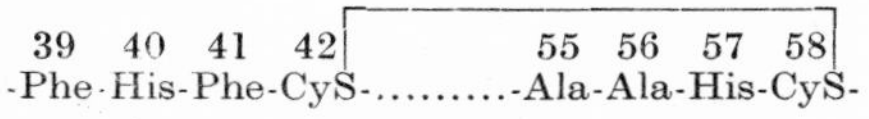

39 40 41 42 ... 55 56 57 58
-Phe-His-Phe-CyS-.........-Ala-Ala-His-CyS-

(CLXXXIV)

28 29 30 31 ... 44 45 46 47
-Tyr-His-Phe-CyS-.........-Ala-Ala-His-CyS-

(CLXXXV)

Photo-oxidation of α-chymotrypsin leads to inactivation and one residue of histidine and one residue of methione are affected. Steady-state and 'all-or-none' assays established that modification of histidine leads to an inactivated enzyme, while modification of methionine leads to a partially active enzyme. Oxidation of α-chymotrypsin with hydrogen peroxide converts the methylthioethyl side chain of one methionyl residue into the sulphoxide; binding of the substrate by the modified enzyme is impaired. Degradative studies have revealed that Met^{192} (CLXXII) is the site of oxidation. Presumably, the side chain of methionine forms part of a non-polar environment which accommodates the hydrophobic side chain of a typical substrate of α-chymotrypsin. It may be of significance that there is a residue of glutamine, which has a hydrophilic side chain, at the corresponding site in trypsin (CLXXIII). The latter enzyme, it will be recalled, catalyses the hydrolysis of derivatives of lysine and arginine, which have a hydrophilic group at the end of a hydrophobic side chain.

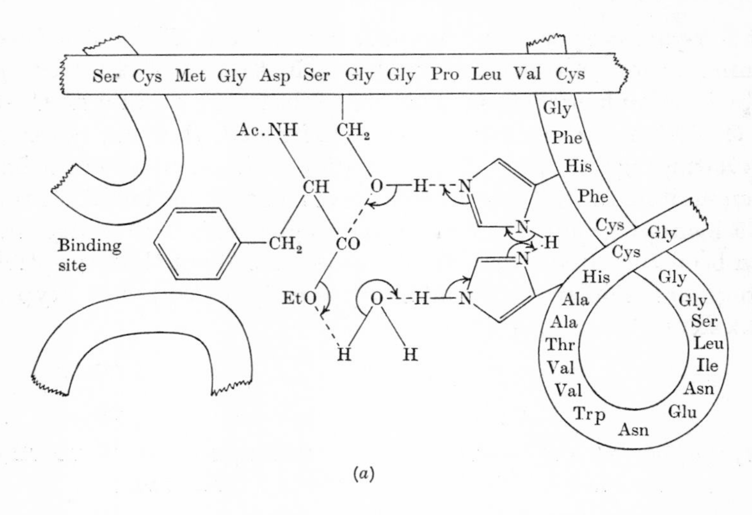

(*a*)

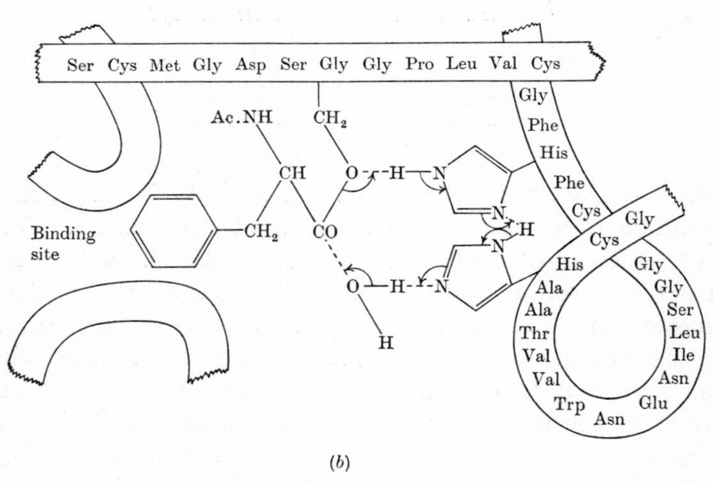

(*b*)

Fig. 6.5. Proposed catalytic mechanism for α-chymotrypsin: (*a*) acylation step, (*b*) deacylation step (Hartley, 1964*a*).

The rate of α-chymotrypsin-catalysed reactions in deuterium oxide is about one-half to one-third of the rate of the corresponding reaction in aqueous solution. This is consistent with a mechanism involving proton transfer in the rate-determining step and suggests that general acid-base catalysis is operative. A proposed mechanism (Hartley, 1964*a*) for the α-chymotrypsin-catalysed hydrolysis of a typical substrate, *N*-acetyl-L-phenylalanine ethyl ester, is illustrated in fig. 6.5. It will be seen that two proximate histidyl residues (presumably His^{40} and His^{57}) are involved, one as the base and the other as its conjugate acid. The high catalytic efficiency of a basic and acidic group in juxtaposition is well known. For example, 2-hydroxypyridine is much more effective than pyridine and phenol for catalysing the mutarotation of tetramethylglucose in benzene, a reaction which is known to require both an acidic and a basic catalyst (Swain & Brown, 1952). The proximity of the acidic hydroxyl group and basic nitrogen atom more than compensates for the fact that 2-hydroxypyridine is a weaker acid then phenol and a weaker base than pyridine. The acylation and deacylation steps in the catalytic mechanism of α-chymotrypsin are reversible by reversing the direction of the electronic shifts. This explains why alcoholysis and hydrolysis compete with one another in aqueous solutions of alcohols. There is, of course, a formal resemblance to the mechanism proposed for ribonuclease with the notable difference that with α-chymotrypsin acylation of enzyme is involved as an intermediate step, while ribonuclease effects the internal acylation of the substrate to give a cyclic phosphate as an intermediate product of reaction. A more detailed discussion of the mechanism of catalysis by α-chymotrypsin has been given by Bender & Kézdy (1964). The mechanism of trypsin catalysis is probably very similar to that of chymotrypsin.

α-Chymotrypsin and trypsin occur in the pancreas as catalytically inert zymogens, chymotrypsinogen and trypsinogen. Both contain single polypeptide chains and their primary structures have been completely determined (Hartley, 1964*b*; Hartley, Brown, Kauffman & Smillie, 1965; Hartley & Kauffman, 1966; Meloun, Kluh, Kostka, Morávek, Prusik, Vanăček, Keil & Šorm, 1966; Walsh & Neurath, 1964; Mikeš, Holeyšovský, Tomášek & Šorm, 1966). Activation of bovine trypsinogen to trypsin is effected by trypsin, i.e. the reaction is autocatalytic. A hexapeptide is split off from the *N*-terminus of the zymogen by rupturing a Lys-Ile bond (CLXXXVI → CLXXXVII + CLXXXVIII)

It is postulated that the amino acids, which constitute the active centre of trypsin, are incorrectly sterically arranged in the zymogen and that activation is accompanied by a conformational

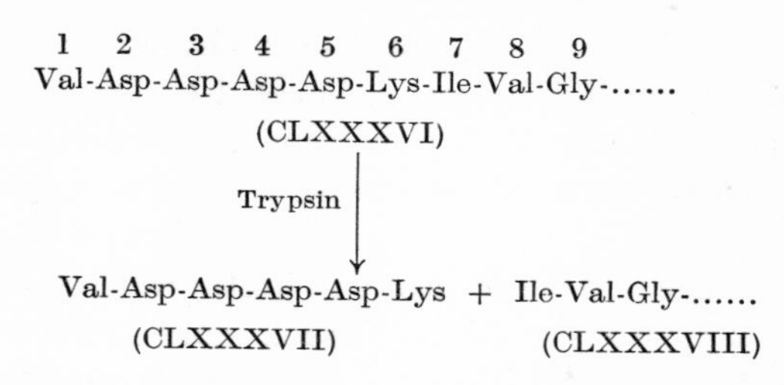

change to establish the active centre. It is noticeable that five of the amino acids in the hexapeptide, which is split off during activation of trypsinogen, could participate in the formation of salt linkages or hydrogen bonds in the zymogen. These might be responsible for maintaining the protein in an inert conformation.

The activation of chymotrypsinogen is rather more complex, since it requires both trypsin and α-chymotrypsin to form

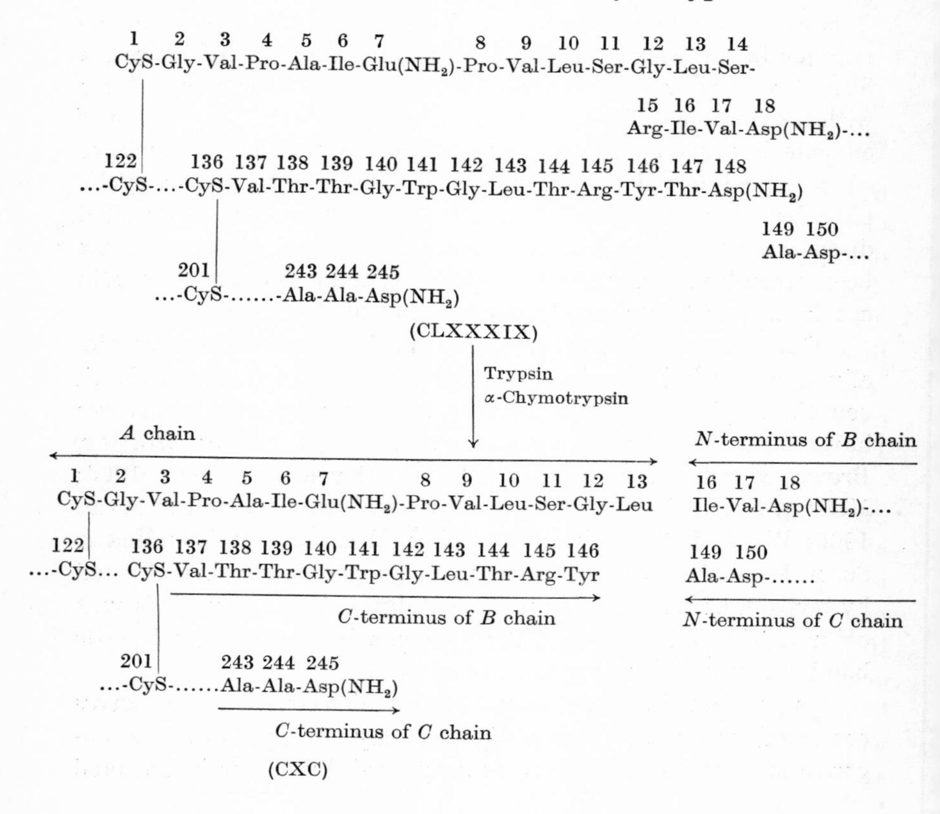

α-chymotrypsin. Two dipeptides, serylarginine and threonylasparagine, are split out from the chain. Trypsin cleaves the $\overset{15}{\text{Arg}}$-$\overset{16}{\text{Ile}}$ bond while chymotrypsin ruptures the $\overset{13}{\text{Leu}}$-$\overset{14}{\text{Ser}}$, $\overset{146}{\text{Tyr}}$-$\overset{147}{\text{Thr}}$ and $\overset{148}{\text{Asp}}(NH_2)$-$\overset{149}{\text{Ala}}$ bonds (CLXXXIX → CXC). The final product, α-chymotrypsin, thus consists of three chains (*A*, *B* and *C*) linked together through disulphide bridges.

Both trypsin and α-chymotrypsin have an *N*-terminal sequence commencing with Ile-Val-. The protonated α-amino group of the *N*-terminal isoleucyl residue in α-chymotrypsin is believed to be important in maintaining the enzyme in a suitable conformation so that it can bind the substrate. Unfortunately, X-ray crystallographic studies have not yet advanced to the stage when it is possible to obtain precise details of the molecular conformations of α-chymotrypsin, trypsin and the corresponding zymogens. The pronounced resemblances exhibited in their primary structures suggest that trypsinogen and chymotrypsinogen may have similar molecular conformations and that at some stage they evolved genetically from the same precursor.

6.13. Cytochrome *c*. Cytochrome *c* forms part of the mitochondrial electron transport chain, which is concerned with the dehydrogenation of succinate and reduced nicotinamide-adenine dinucleotide and the eventual formation of water. Cytochrome *c* is a haemoprotein and the prosthetic group (fig. 6.6) is covalently linked to the protein through the side chains of two cysteinyl residues. Electron transport is effected by reversible reduction and oxidation of the iron in the prosthetic group between the Fe^{2+} and Fe^{3+} states.

The sequence of amino acids has been determined for samples of cytochrome *c* from a number of species. Cytochrome *c* from almost all vertebrates has 104 amino acids and the *N*-terminal amino acid is acetylated. Cytochrome *c* from some non-vertebrates contains more than 104 residues; when the sequences are compared with those from vertebrates, it appears that the additional residues occur at the *N*-terminus. It is convenient for sequence comparison to start numbering the residues from the *N*-acetylated amino acid, which is the normal *N*-terminus of cytochrome *c* from vertebrates, and to number any additional residues backwards from this point (fig. 6.7). Cytochrome *c* is rich in lysine and this amino acid tends to occur in nodes along the single polypeptide chain. Similar clustering of hydrophobic amino acids is found.

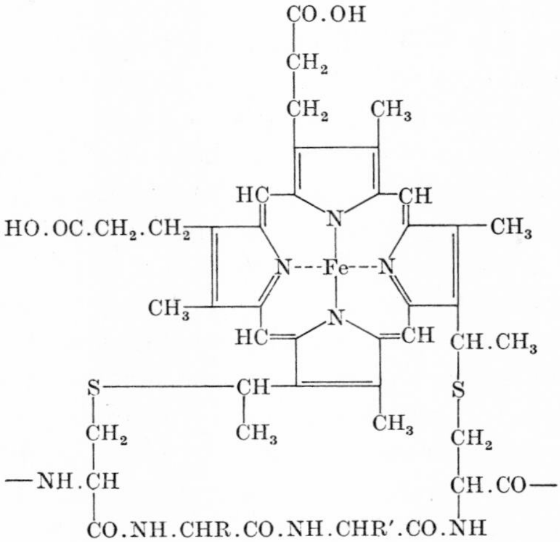

CO.OH
CH_2
CH_2 CH_3
HC— —CH
$HO.OC.CH_2.CH_2$— N —CH_3
N- - -Fe- - -N
CH_3— N
HC= =CH $CH.CH_3$
S———CH— S
CH_2 CH_3 CH_3 CH_2
—NH.CH CH.CO—
CO.NH.CHR.CO.NH.CHR′.CO.NH

Fig. 6.6. The prosthetic group of cytochrome *c* showing the linkage to Cys[14] and Cys [17] and the intervening two amino acid residues.

−5	−4	−3	−2	−1	1	2	3	4	5	6	7	8
							Ser					
			Ser			Ser	Ala	Lys				Thr
	Gly	Val	Pro			$Asp(NH_2)$	Ile	Ala	$Asp(NH_2)$		Ala	$Asp(NH_2)$
Thr	Glu	Phe	Lys	Ala	GLY	-ASP	-VAL	-GLU	-LYS	-GLY	-LYS	-LYS

9	10	11	12	13	14	15	16	17	18	19	20
					HAEM						
		Lys									
Thr		Thr	Thr			Ala	Glu				
Leu		Val	$Glu(NH_2)$	Arg		Glu	Leu			Gly	Glu
ILE	-PHE	-ILE	-MET	-LYS	-CYS	-SER	$-GLU((NH_2)$	-CYS	-HIS	-THR	-VAL-

21	22	23	24	25	26	27	28	29	30
	Gly								
	Ala			Thr			Ile		
Gly	$Asp(NH_2)$	$Asp(NH_2)$	Leu	Pro	$Glu(NH_2)$		Val		
GLU	-LYS	-GLY	-GLY	-LYS	-HIS	-LYS	-THR	-GLY	-PRO-

31	32	33	34	35	36	37	38	39	40	41
		Trp		Phe	Ile					
Ala		$Asp(NH_2)$		Ile	Tyr			His	Ser	
$ASP(NH_2)$	-LEU	-HIS	-GLY	-LEU	-PHE	-GLY	-ARG	-LYS	-THR	-GLY-

Position	42	43	44	45	46	47	48	49	50	51
Variants	Ser	Val	Asp Val Glu Glu(NH_2)		Phe	Ala Thr		Ser	Glu Asp Asp(NH_2)	
Human	GLU(NH_2)	ALA	PRO	GLY	TYR	SER	TYR	THR	ALA	ALA

Position	52	53	54	55	56	57	58	59	60	61
Variants		Ile	Ala Ser Lys Glu(NH_2)		Asp(NH_2)	Val	Thr Val Leu		Lys Asp Asp(NH_2)	Asp Asp(NH_2)
Human	ASP(NH_2)	LYS	ASP(NH_2)	LYS	GLY	ILE	ILE	TRP	GLY	GLU

Position	62	63	64	65	66	67	68	69	70	71	72
Variants	Glu Asp(NH_2)	Asp(NH_2)	Met	Ser Phe				Thr			
Human	ASP	THR	LEU	MET	GLU	TYR	LEU	GLU	ASP(NH_2)	PRO	LYS

Position	73	74	75	76	77	78	79	80	81	82	83	84	85	86
Variants									Ala Val		Thr Ala Gly		Leu	Ser
Human	LYS	TYR	ILE	PRO	GLY	THR	LYS	MET	ILE	PHE	VAL	GLY	ILE	LYS

Position	87	88	89	90	91	92	93	94	95	96	97
Variants		Ala Thr Asp Glu	Ala Lys Thr Ser Gly Asp(NH_2) Asp	Asp		Asp(NH_2) Glu(NH_2) Glu Thr Val	Asp(NH_2)	Ile	Val	Thr	Phe
Human	LYS	LYS	GLU	GLU	ARG	ALA	ASP	LEU	ILE	ALA	TYR

Position	98	99	100	101	102	103	104
Variants	Met		Ser Glu Asp	Lys Ser		Lys CySH Ser Ala	Lys Ala
Human	LEU	LYS	LYS	ALA	THR	ASP(NH_2)	GLU

Fig. 6.7. Primary structures of cytochrome *c* from 18 species: man, horse, *Macacus mulatta*, pig, cow, sheep, whale, rabbit, dog, kangaroo (*Macropus canguru*), chicken, turkey, rattlesnake (*Crotalus adamanteus*), tuna fish, snapping turtle (*Chelydra serpentina*), moth (*Samia cynthia*), baker's yeast and *Neurospora crassa*. The primary structure of human cytochrome *c* is printed in capitals. The amino acid residues given at each locus include all the variations found in the above 18 species. Notice the position of the prosthetic group and the almost invariant sequence (66–80 inclusive). In most species Gly[1] is *N*-acetylated; in *Samia cynthia*, baker's yeast and *Neurospora crassa*, the *N*-acetyl group is replaced by additional amino acid residues.

When the primary structures of cytochrome *c* from various sources are compared (fig. 6.7), a remarkable degree of correspondence is found. Particularly noticeable features are (*a*) the position of the cysteinyl residues carrying the prosthetic group and the adjacent histidyl residue at position 18, (*b*) the constancy of the sequence 66–80, (*c*) the tendency for many of the sequence variations to involve amino acids of similar types, e.g. Phe and Tyr, Ser and Thr, Leu and Ile.

If the primary structures of cytochrome *c* from any two species are compared, it is found that the number of differences increases with the length of time since the two lines are believed to have diverged. Among mammals, for example, cytochrome *c* is identical from pig, cow and sheep, while cytochrome *c* from man and Rhesus monkey (*Macacus mulatta*) differ at only one locus. The maximum number of differences between mammals so far found is twelve (man: horse). When the sequences of cytochrome *c* from a group of mammals is compared with that from chicken (table 6.1) there is an average of 10·5 variant residues. If it is assumed that mammalian and avian lines diverged 280 million years ago, one amino acid has been changed every 26·7 million years (cf. Margoliash & Schejter, 1966). Using this figure and the average number of differences between the primary structures of cytochrome *c* from mammals and from lower phyla and classes, it can be calculated when the various lines diverged. The calculated time of divergence is likely to be an underestimate, since more than one mutation may have occurred at the same site and will not therefore be counted.

It is obvious that the prosthetic group is essential for the activity of cytochrome *c*, but the role of the remainder of the protein is less clear. Two groups are presumably additional ligands for the iron atom. One of these is almost certainly histidine, since photo-oxidation of cytochrome *c* destroyed its activity and oxidized one of the histidyl residues at a parallel rate. Since His^{18} is the only invariant histidyl residue, it is probable that the imidazole ring of this amino acid forms one of the two additional ligands. Identification of the other ligand is much more nebulous. An ϵ-amino group of a lysyl residue or the sulphur atom in the side chain of a methionyl residue have been suggested as possible candidates. For a detailed account of the prosthetic group of cytochrome *c*, the review by Margoliash & Schejter (1966) should be consulted.

TABLE 6.1. *Evolution of Cytochromes c*

Species comparison	Variant residues Number	Average	Divergence of lines in millions of years
Rabbit : Chicken	8		
Pig, cow, sheep : Chicken	9		
Whale : Chicken	9		
Dog : Chicken	10		
Horse : Chicken	11	10·5	280 (assumed)
Kangaroo : Chicken	12		
Monkey : Chicken	12		
Man : Chicken	13		
Rabbit : Tuna	17		
Pig, cow, sheep : Tuna	17		
Whale : Tuna	17		
Dog : Tuna	18		
Horse : Tuna	19	18·7	490 (calculated)
Kangaroo : Tuna	20		
Man : Tuna	21		
Monkey : Tuna	21		
Chicken : Tuna	17		
Rabbit : Moth	26		
Pig, cow, sheep : Moth	27		
Whale : Moth	27		
Dog : Moth	26		
Horse : Moth	29		
Kangaroo : Moth	29	28·5	760 (calculated)
Monkey : Moth	30		
Man : Moth	31		
Chicken : Moth	27		
Tuna : Moth	33		

There is little definite information concerning the role of other amino acids in the protein. Trifluoroacetylation of ϵ-amino groups caused loss of activity which was restored when the blocking groups were removed. On the other hand, complete guanidination of ϵ-amino groups can be effected without loss of activity. Alkylation with iodoacetic acid led initially to reaction with Met^{65} and His^{33} without loss of activity, but further alkylation effected modification of Met^{80} with complete loss of activity. It is perhaps significant that Met^{80} is at the end of the largest invariant sequence in cytochrome *c*. It is possible that much of this region is essential for biological activity.

6.14. Epilogue. It should be clear now that several techniques such as chemical synthesis of analogues and simple models, X-ray crystallography, modification studies, kinetic studies with enzymes, and sequence determination of homologous proteins from different species, are all capable of adding to our knowledge concerning the mechanism of action of biologically active polypeptides. No technique by itself is likely to prove adequate and there is a great need for additional methods. The prizes are considerable. Biochemistry is developing from the stage of finding out what goes on in a living cell or organism to the more difficult and perhaps intellectually satisfying step of explaining how these processes occur. Not all the prizes are likely to go to biochemists; the elucidation of biochemical reactions at the molecular level is as much the province of the chemist and perhaps the physicist as it is of the biochemist. Moreover, if we could understand biochemical reactions at the molecular level, we would be better able to design and synthesize molecular spanners (drugs) to throw into the works in order to modify selectively the kinetic perturbations which are universally manifested in disease and ageing. 'Anti-enzyme guided missiles' might selectively inhibit unwanted or overactive enzymes, while synthetic model-enzymes might replenish or replace those which are deficient or missing.

Bibliography

Adair, G. S. (1961). In *Analytical Methods of Protein Chemistry*, vol. 3, p. 23, Ed. by P. Alexander and R. J. Block. Oxford: Pergamon Press, Ltd.

Arnstein, H. R. V. (1965). *Brit. Med. Bull.* **21**, 217.

Baker, B. R. (1964). *J. Pharm. Sci.* **53**, 347.

Bender, M. L. & Kézdy, F. J. (1964) *J. Amer. chem. Soc.* **86**, 3704.

Benesch, R. & Benesch, R. E. (1962). *Methods Biochem. Anal.* **10**, 43.

Bennett, J. C. & Dreyer, W. J. (1964). *Ann. Rev. Biochem.* **33**, 205.

Bergman, M. & Zervas, L. (1932). *Ber.* **65**, 1192.

Bier, M. (1959). *Electrophoresis: Theory, Methods and Applications.* New York: Academic Press, Inc.

Claesson, S. & Moring-Claesson, I. (1961). In *Analytical Methods of Protein Chemistry*, vol. 3, p. 119, Ed. by P. Alexander and R. J. Block. Oxford: Pergamon Press, Ltd.

Cohn, E. J. & Edsall, J. T. (1943). *Proteins, Amino Acids and Peptides*, p. 370. New York: Reinhold Publishing Corporation.

Cohn, E. J., Strong, L. E., Hughes, W. L., Mulford, D. J., Ashworth, J. N., Melin, M. & Taylor, H. L. (1946). *J. Amer. chem. Soc.*, **68**, 459.

Craig, L. C. (1960*a*). In *Analytical Methods of Protein Chemistry*, vol. 1, p. 103, Ed. by P. Alexander and R. J. Block, Oxford: Pergamon Press, Ltd.

Craig, L. C. (1960*b*). In *Analytical Methods of Protein Chemistry*, vol. 1, p. 121, Ed. by P. Alexander and R. J. Block. Oxford: Pergamon Press, Ltd.

Craig, L. C. (1962). In *Comprehensive Biochemistry*, vol. 4, p. 1, Ed. by M. Florkin and E. H. Stotz. Amsterdam: Elsevier Publishing Co.

Crick, F. H. C. (1963). In *Progress in Nucleic Acid Research*, vol. 1, p. 164, Ed. by J. N. Davidson and W. E. Cohn. New York: Academic Press, Inc.

Crick, F. H. C. (1966). *Scient. Amer.* **215**, 55.

Deavin, A., Mathias, A. P. & Rabin, B. R. (1966). *Nature* **211**, 252.

Dickerson, R. E. (1964). In *The Proteins*, 2nd ed., vol. 2, p. 603. Ed. by H. Neurath. New York: Academic Press, Inc.

Djerassi, C. (1960). *Optical Rotatory Dispersion.* New York: McGraw-Hill Book Co., Inc.

Edman, P. & Begg, G. (1967). *Europ. J. Biochem.* **1**, 80.

Edsall, J. T., Flory, P. J., Kendrew, J. C., Liquori, A. M., Némethy, G., Ramachandran, G. N. & Scheraga, H. A. (1966). *Biopolymers.* **4**, 121.

Goodman, M. & Kenner, G. W. (1957). *Adv. Protein Chem.* **12**, 465.

Greenstein, J. P. & Winitz, M. (1961). *Chemistry of the Amino Acids*, vol. 2, p. 763. New York: J. Wiley and Sons, Inc.

Harding, J. J. (1965). *Adv. Protein Chem.* **20**, 109.

Hartley, B. S. (1964*a*). In *Structure and Activity of Enzymes*, p. 47, Ed. by T. W. Goodwin, J. I. Harris and B. S. Hartley. London: Academic Press (London), Ltd.

Hartley, B. S. (1964*b*). *Nature* **201**, 1284.

Hartley, B. S., Brown, J. R., Kauffman, D. L. & Smillie, L. B. (1965). *Nature* **207**, 1157.

Hartley, B. S. & Kauffman, D. L. (1966). *Biochem. J.* 101, 209.

Hill, R. L. (1965). *Adv. Protein Chem.* **20**, 37.

Hofman, K. & Katsoyannis, P. G. (1963). In *The Proteins*, 2nd ed., vol. 1, p. 53, Ed. by H. Neurath. New York: Academic Press, Inc.

James, A. T. & Morris, L. J. (1964). *New Biochemical Separations.* London: Van Nostrand Co., Ltd.

Karmen, A. & Saroff, H. A. (1964). In *New Biochemical Separations*, p. 81, Ed. by A. T. James and L. J. Morris. London: Van Nostrand Co., Ltd.

Kartha, G., Bello, J. & Harker, D. (1967). *Nature* **213**, 862.

Kenchington, A. W. (1960). In *Analytical Methods of Protein Chemistry*, vol. 2, p. 353, Ed. by P. Alexander and R. J. Block. Oxford: Pergamon Press, Ltd.

Keosian, J. (1964). *The Origin of Life.* London: Chapman and Hall, Ltd.

Kurzer, F. & Douraghi-Zadeh, K. (1967). *J. org. Chem.* **67**, 107.

Lederer, E. & Lederer, M. (1957). *Chromatography.* Amsterdam: Elsevier Publishing Co.

Light, A. & Smith, E. L. (1963). In *The Proteins*, 2nd ed., vol. 1, p. 1, Ed. by H. Neurath. New York: Academic Press, Inc.

Margoliash, E. & Schejter, A. (1966). *Adv. Protein Chem.* **21**, 113.

Mathias, A. P., Deavin, A. & Rabin, B. R. (1964). In *Structure and Activity of Enzymes*, p. 19. Ed. by T. W. Goodwin, J. I. Harris and B. S. Hartley. London: Academic Press (London) Ltd.

Meloun, B., Kluh, I., Kostka, V., Morávek, L., Prusik, Z., Vanăček, J., Keil, B. & Šorm, F. (1966). *Biochim. Biophys. Acta* **130**, 543.

Merrifield, R. B. (1965). *Science* **150**, 178.

Mikeš, O., Holeyšovský, V. Tomášek, V. & Šorm, F. (1966). *Biochem. Biophys. Res. Commun.* **24**, 346.

Miller, S. L. (1955). *J. Amer. chem. Soc.* **77**, 2351.

Morris, C. J. O. R. & Morris, P. (1963). *Separation Methods in Biochemistry.* London: Sir Isaac Pitman and Sons, Ltd.

Oparin, A. I. (1965). *Adv. Enzymol.* **27**, 347.
Partridge, S. M. (1962). *Adv. Protein Chem.* **17**, 227.
Pattee, H. H. (1965). *Adv. Enzymol.* **27**, 381.
Peterson, E. A. & Sober, H. A. (1960). In *Analytical Methods of Protein Chemistry*, vol. 1, p. 88, Ed. by P. Alexander and R. J. Block. Oxford: Pergamon Press, Ltd.
Russell, D. W. (1966). *Quart. Rev.* **20**, 559.
Sanger, F. (1955). *Bull. Soc. Chim. biol.* **37**, 23.
Schröder, E. & Lübke, K. (1965). *The Peptides*, vol. 1. New York: Academic Press, Inc.
Schröder, E. & Lübke, K. (1966). *The Peptides*, vol. 2. New York: Academic Press, Inc.
Schwyzer, R., Costopanagiotis, A. & Sieber, P. (1963). *Helv. Chim. Acta*, **46**, 870.
Smith, E. L. & Hill, R. L. (1960). In *The Enzymes*, vol. 4, p. 37, Ed. by P. D. Boyer, H. Lardy and K. Myrbäck. New York: Academic Press, Inc.
Stewart, J. M. & Woolley, D. W. (1965). *Nature* **207**, 1160.
Stretton, A. O. W. (1965). *Brit. Med. Bull.* **21**, 229.
Svensson, H. & Thompson, T. E. (1961). In *Analytical Methods of Protein Chemistry*, vol. 3, p. 57, Ed. by P. Alexander and R. J. Block. Oxford: Pergamon Press, Ltd.
Swain, C. G. & Brown, J. F. (1952). *J. Amer. chem. Soc.* **74**, 2534, 2538.
Tanford, C. (1962). *Adv. Protein Chem.* **17**, 69.
Thompson, E. O. P. (1960). *Adv. org. Chem.* **1**, 149.
Ulbricht, T. L. V. (1959). *Quart. Rev.* **13**, 48.
Urnes, P. & Doty, P. (1961). *Adv. Protein Chem.* **16**, 401.
Walsh, K. A. & Neurath, H. (1964). *Proc. Nat. Acad. Sci.* **52**, 884.
Watson, J. D. (1965). *Molecular Biology of the Gene*. New York: W. A. Benjamin, Inc.
Wetlaufer, D. B. (1962). *Adv. Protein Chem.* **17**, 303.
Witkop, B. (1961). *Adv. Protein Chem.* **16**, 221.
Yamagata, Y. (1966). *J. Theor. Biol.* **11**, 495.

Index